YouTube Ideas for Creatives & Entertainers

Over 200 Video Ideas to Unleash Your Creativity and Craft Captivating Content

Ben Davis

Contents

The Creatives & Entertainers Alcove

Greetings, fellow architects of amusement and connoisseurs of recorded media! Welcome to a book where we'll be embarking on a journey into the very heart of digital imagination, exploring a myriad of video ideas you can create within the categories of creative and entertaining YouTube content.

As a dedicated video maker myself, I understand the constant quest for new ways to capture and enthrall your audience, because I share that same drive. Within the pages of this book, I offer my curated list of video ideas and share my process of tweaking them to align with your own unique style.

We'll talk about everything from the wildest and most popular competitions to the most enjoyable "try not to laugh" challenges for the entertainers in the house. Every idea within these two categories will elevate your channel to the next level and encourage subscribers to hit that "Like" button.

I encourage you to explore the concepts most aligned with your current niche and test their engagement with your audience. At the same time, you can discover whether the ideas resonate with your creative instincts. From there, you can decide whether you want to record more of the same type of video or broaden your horizons and try another.

Not only will this guide help you explore both new and timeless niche concepts, it will also teach you how to effortlessly craft your own original content. Once you've mastered the art of bringing any idea to life in your own unique style, the possibilities will become limitless. Soon, you'll find yourself creating without the need for a guide. With more than 150 content ideas in this book to get you started, you only need to choose one. Then, you'll be able to plan, record, and evaluate the video before you move on to the next idea.

When you find a video concept you like, set the book aside and keep making this type of video until you feel that you need a fresh idea. Whether you are a seasoned YouTuber or new to the ranks, you can use this guide to help you stay on track. This way, you'll be able to consistently and constantly build and reinforce your content creation skills.

These ideas are not simply limited to YouTube, either. You will find concepts that are easy to carry over to TikTok, Instagram, and any other format that you might want to use to reach your perfect viewers.

As this book is a tool for content creation, use it as such—not only to think of new video ideas, but also as a way to help you plan content strategies ahead of time. To do this, simply flick through the array of ideas in the category of your choice and write down inspiration in the space provided under each idea. This way, you are free to generate your own ideas without being limited by too much structure.

For instance, under "DIY Hacks-Productive Life Hacks," you could note down things you do in your own house that you've recently found useful. Here are examples of notes you could use for this section:

- Use foil for cleaning sink (makes it shiny).
- Repurpose oven foil after dinner (makes it sustainable).
- Include funny element (the sink smells like roast, but it looks great).

This is your basis for planning content, but you can keep it all in one place in this book until you are ready to use it.

The content industry has played a significant role in keeping the world's collective serotonin and dopamine levels high during tough times. This is because it has allowed for personal artistic expressions to be shared through social media.

You do not need to possess an X-factor to immerse yourself in this newer artistic world; just be yourself and to follow your passions. Each of us already has a unique blend of artistic flair and wit that will connect with a particular audience. As an artist, all you need to do is bring yourself into the creation process.

It doesn't matter if you prefer making creative videos or entertaining audiences—only that you inject your personality into the mix. While we explore the core of digital creation, idea-generation fatigue can become a real struggle. This book will also help you with that, effortlessly committing to crafting content that brings laughter, joy, and even introspection by keeping it simple.

Ideas are like fireflies; they spark and are beautiful, but if you don't capture them, they will fly off. As such, your job is to meticulously jot down your ideas as they come. Obviously, this book only has room for a few bullet points under each idea that I've laid out. So, in the theme of keeping it simple, you can flesh out ideas in a separate workbook once you've decided on the video that you want to plan.

Designing and planning narratives are outside of the scope of this book, so stick to idea generation here.

As creators and entertainers, don't feel like you need to limit yourself to one category, either. You can define yourself by your ability to be versatile. Simply confining yourself to one single idea, one format, or one constant formula is an incredible limitation and one that artists do not necessarily need to adhere to.

Instead, you could elegantly navigate through a myriad of different worlds and transitions, and this book can help you do it. As a first step, you could spend some time reading and jot down ideas, even if the categories don't seem to mesh well with your current theme. You might be surprised by how your thoughts seamlessly link ideas from the realm of DIY crafts to the explosive competitive fields of challenges.

This sounds like gold viewing to me, and that's exactly what's needed to ensure that your content will remain dynamic and engaging for a diverse audience.

Even if it seems as if everything's been done before, you will find that when you go to make your video, the end result will be different from what is currently out there. Any video maker knows that you start with an idea, and it grows and changes as you push yourself toward the final version.

In Part One of this book, we'll leap into the world of entertainment to inject a bit of adrenaline and excitement into your content. In this arena, it's about discovering how far you can push yourself and keeping your audience on the edge of their seats.

You'll be introduced to challenge videos with a twist, which will help you to transform the ordinary into the extraordinary. You'll learn to blend old with new into a beautifully fresh symphony of content. Whether it's tackling blindfold challenges, embracing a day—or perhaps more—without technology, or taking a deep-dive into the realm of the virtual, every twist and turn will bring an additional layer of intrigue and laughter.

We'll look at virtual odysseys through travel and exploration, whisking them away to the far-flung corners of the Earth. From their homes, viewers are immersed in diverse cultures, landscapes, and quests to capture the essence of travel. These digital travelogues are meticulously documented experiences of firsthand experiences, discoveries, and awesome pictures of each destination's unique beauty.

This becomes an inspiring gateway for those who boast the spirit of wanderlust, sparking a sense of curiosity and appreciation for the unfamiliar. Whether uncovering hidden gems in distant cities, embarking on nature retreats, or sailing the world's oceans, creatives have endless possibilities for transporting their audiences into new realms of discovery.

Another famous genre across social media and other entertainment platforms is that of reactions. This is a diverse world of emotional roller coasters, shock, awe, disgust, and perhaps even fear. From watching the strangest of TikToks to discussing global politics, from nostalgic trips to 2000s reality shows to lamenting the loss of great advertisements, reaction channels have the opportunity to cover it all.

Some reactors offer celebrity impersonations, stepping into the shoes of others with impressive and effective ease. Some may only tease, while others push the envelope until it verges on the absurd. This content often grasps younger audiences, who find hilarity in the overdramatic.

And you don't have to spend hours on the perfect make-up and hair for impersonations to go viral; a towel and a good joke will reach millions in just a few hours. There's something about the simple style and the uniqueness of this kind of comedy that leads to both relatable *and* entertaining content, both easy to watch and to enjoy.

Then there are nostalgia trips, where both reactors and viewers join one another in reminiscing about the trends of the past, about childhood favorites, and about personal milestones. This content evokes the good, old days, blending sentimentality and humor into a shared experience that has the power to transcend the boundaries of generations.

You offer a bridge to your audience for shared experiences and connections that they might not otherwise have gotten. It is a unique mark to leave in the digital world, a relatable and humorous exploration of the human psyche that evolves alongside our technology and culture.

In Part Two of this book, I speak to the creatives, those video-makers with a passion for exploring artistic expression through various fascinating means. Perhaps you've never been able to keep a plant alive, and you'd like to challenge yourself to create a garden, where the lack of a green thumb combined with comedic genius can turn gardening into a spectacle. Or perhaps you'd like to try and renovate your bedroom or build a complex Swedish piece of furniture without any English instructions. Or perhaps you are a creative maestro, an artist with a voice and a wish to share their world with others.

Many have carved their way into this niche through the metamorphosis of everyday tasks into delightful adventures. Not only does it offer your audience a chance to find new insights and valuable hacks, but the truly talented creator can bring a healthy dose of laughter to the table with it. You can choose from a variety of subcategories, from hacks to how-to guides all the way to quirky crafts to show off your unique thinking and creativity. Perhaps you are a true genius and can emerge as one of the top creators in the genre.

Creatives truly have the world laid open in front of them. Some may take on the role of the fashionista in costume challenges, where clothing defies convention. Others may prefer the unpredictability of extreme weather challenges, where they face the incredible forces of nature and the elements head-on, turning storms into adventure.

Then there's education and learning, a beacon of knowledge in a vast sea of content online. This niche is an opportunity to soar above traditional educational boundaries with informative and engaging content. These mentors captivate audiences globally with how-to guides, tutorials, experiments, and challenges. In so doing, they have managed to transform learning into something that's not only interactive and educational, but enjoyable as well.

Virtual mentors can skillfully guide their viewers through almost any subject with clarity and a clear passion for their work. This has led to the creation of the well-known adage "if you want to learn something new, just ask YouTube." This is a testament to the sheer power of the digital to democratize education by creating accessible learning that can build lifelong journeys of discovery for viewers of all ages.

In essence, video-makers across all niches showcase their expressions as celebrations of human creativity, humor, and the ever-evolving possibilities of the internet. They are architects of digital experience, weaving stories that resonate with the very core of the human spirit. They shape our modern cultural landscape in profound ways by thriving on innovation, connection, and the boundless potential of the human imagination.

Be kind, have fun, and stay authentic to yourself, then go out there and embark on this exciting journey of brilliant video-making.

Part One

YouTube Ideas for the Entertainers

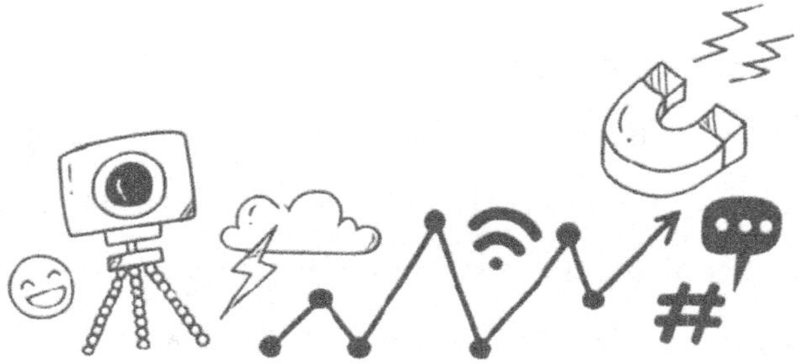

Try-Not-to-Laugh Challenges

Are you a brave, stoic warrior, who cannot be moved by any emotions? To truly succeed in a try-not-to-laugh challenge, you have to be. But really, the entertainment of these challenges isn't in being constantly stony-faced. In fact, the secret lies in *losing* the challenge dramatically while trying your utmost best to keep a straight face.

At the same time, you should be encouraging viewers to battle it out with you, testing their own resolve when it comes to poker-facing through the funniest videos on the internet. This offers an interactive element that is unique to this niche, increasing engagement and fostering a sense of relatability and friendship with you as the video maker.

This challenge has only one objective: don't laugh. It's your willpower against humor, whether that's cat videos, unexpected fails, dad jokes, or really any other kind of silliness that can be thrown at you from the weird world wide web.

Can you keep your composure in the face of all of this chaos? That's the real question here, and the answer is usually a resounding "no!" No matter how many times you lose a try-not-to-laugh challenge, though, it's guaranteed to bring your audience back for more.

The following 11 suggestions should tickle your funny bone.

Animal Antics

Set up your video to record yourself while you watch these videos and stream a perfectly curated collection in the background. Challenge yourself to keep a poker face while your audience watches, enjoying both the hilarious videos and your attempts to stifle your own laughter. Seek out the best funny-animal videos out there that are sure to tickle your funny bone, like a dog accidentally smashing into a pile of fall leaves, or perhaps a cat longing to capture a bird on the other side of a window.

You might even discover amusement in observing nature and wildlife, such as watching a perplexed bear attempt to climb a tree two sizes too small, or a lion rolling about gleefully in the sand. From stories of the smallest mouse to the greatest elephant, there are millions of funny animal videos to choose from for your next challenge that are certain to truly entertain your audience.

2. Prank Fails

Everyone enjoys watching creative pranks; it's even more entertaining when they go hilariously wrong. Take advantage of this by collecting videos that you find particularly amusing due to their failed attempts and challenge your audience not to laugh as you play them.

Search for clips in which the prankster accidentally triggers their own trap or encounters some other misfortune before they can pull off a successful practical joke. These types of videos often elicit unexpected reactions from participants, so encourage your audience to comment on them and let you know why they laughed in your comment section. This will boost engagement, help foster your community, and improve your video's visibility in the algorithm, leading to more views.

3. Epic Dad Jokes

These are classics in the world of cheesy humor, with jokes ranging from groan-worthy to sheer knee-slapping hilarity. Search for all the best dad-joke challenges, where dads face off against one another with the best comedy in their arsenals, or a simple series of the highest-quality cheesy jokes that you can find on the internet. From here, ask your audience to join you as you record yourself watching these videos and stream them as you view.

At the beginning, it might be easy to remain stoic, but as these jokes intensify, resistance becomes increasingly futile. If you are looking for a unique way to level up your anti-laughing muscles, dad jokes are certainly the way to go. Ask your audience to add to the hilarity by sharing their own jokes in the comment section, boosting engagement and leveling up your position in the algorithm.

____

____

____

____

4. Baby Laughter

Of all the types of infectious laughter in the world, baby laughter is some of the very strongest. Innocent young ones will giggle at seemingly anything, from their father making faces to a balloon floating across a park. Their joy and happiness can warm the hearts of even the stoniest creators, and you'll often find yourself laughing along without thinking. It's truly difficult to keep a straight face when watching the littlest of people enjoying their very first years of life—both for you and your audience.

Therefore, it is always a fun and fresh idea to line up a collection of your favorite baby-laughter videos and challenge your viewers not to laugh along. Set up your recording and allow your audience to join in watching while you attempt to avoid the infectious nature of infant laughter. Suggest discussions about laughter in the comments section, and ask viewers to share who in their lives they believe has the most infectious laugh.

5. News Bloopers

The news media is often thought of as a serious world full of scary headlines and deep debates. But even the best news anchors are, after all, only human, and with many of their segments broadcast live, there are bound to be hundreds of new bloopers every day. News stations from all over the world in all kinds of languages can offer hilarious moments, mistakes, and unexpected incidents that are sure to tickle anyone's funny bone.

Research the funniest moments caught live on the news cameras and showcase this to your audience. Attempt not to laugh as you watch weather mishaps and fumbled headlines while encouraging viewers to do the same. Ask about the most famous news bloopers in their home country. This encourages viewers from across the globe to engage with your content and shows your enthusiasm for connecting with them. More comments will also increase your chances of a viral video, since engagement is the key.

6. Funny Kid Moments

Children famously have no filters, and this kind of compilation uses this to its advantage. Kids will test their boundaries in ways that adults could not even imagine, saying ridiculous things or attempting athletic feats that will leave viewers in stitches.

Try to find the most unique videos that you can in this genre, as it's one that's widely viewed by reactors. Ensure that your curated collection is unlike anything else that your audience has seen before, and add a bit of your own flavor into the mix. Set up your recording, let the videos play, and challenge your viewers not to burst into laughter before you do. Ask them to mention the silly things they said as children in the comments. This is a relatable and easy discussion to have that will surely foster engagement and a hilarious repartee as commenters share their memories with one another.

7. Random Dance Challenges

Platforms like TikTok, Facebook, and Instagram are filled with viral attempts at the newest dances, often with disastrous and hilarious consequences. But these are not the only places where you might be able to source the funniest dances known to man. You can look at game shows and talent shows, and perhaps even reality TV to find failed—or even successful—attempts at famous or infamous dances.

This will keep you on top of current trends and ensure your content is fresh and new. While recording yourself watching these videos and letting your viewers join you, you can add memes and edits in between in keeping with the dancing theme. This will give your video an additional sense of rhythm and excitement, making sure that something interesting is constantly happening.

8. Weird Commercials

The world of advertising is often a weird and wonderful place. Across the globe, marketing agencies attempt to create something for their clients that will stand out in the market and draw in audiences. This often leads to bizarre creations that will leave you howling with laughter, and which certainly offer a unique angle to the try-not-to-laugh challenges that we know and love.

With this challenge, your aim is to find the most bizarre and strange commercials that you can, whether they're vintage infomercials from the US or modern detergent ads from India. Search for commercials from all over the world and curate the strangest collection on the market. Then, record yourself while streaming them, and allow your audience to witness the odd creativity that marketers the world over have come up with. Encourage viewers to share the most infamous commercials from their own country or region in the comments, too, to ensure that you continue to gain maximum engagement.

9. Unexpected Celebrity Moments

Just like us, celebrities are human, too. Their every waking moment is filmed and captured on camera, and whenever something goes hilariously wrong, recordings of it will flood social and traditional media very quickly. Use this to your advantage by finding the most unexpected celebrity moments and collecting them in a hilarious compilation.

Then, sit down with your equipment and challenge both yourself and your viewers not to laugh at the antics of Hollywood's most famous people. Ask about your audience's strangest encounters with celebrities and encourage discussions in the comments. This will foster connection and relatability, which will strengthen your community—and increase engagement.

10. Comedic Mashups

Many YouTubers thrive in the niche of comedic skits and humorous stylings, and it's often a great idea to connect with them. Create mashups of their videos to challenge yourself not to laugh, and use popular content like Lele Pons's sketches, with their high energy and wide fan base to keep your audience hooked.

Covering other YouTubers, particularly highly popular ones, will increase your views and the engagement on your channel. You can even collaborate with other channels in creating your mashups, and record yourselves reacting to these compilations together. Connecting in this way will bring their viewers to your channel, and your viewers to theirs—a true triumph for all involved.

11. Television Shows

The world has an insane variety of strange reality TV shows, game shows, and talent shows that can offer a wealth of content when it comes to attempting not to laugh. From strange addictions to ridiculous competitions, you can find nearly anything you can imagine to *not* snicker at. Through this lens, you can introduce your audience to brand-new content and strange realms that they never would have found on their own.

Once you've curated your collection of odd television shows, settle in to record your reactions to their most insane moments. Ask your audience to share riveting tales of their favorite shows in the comments while you attempt to hold back your own laughter. This will connect you and your audience in a shared consciousness and offer them a strong chance to relate to you.

Try-not-to-laugh challenges are incredibly popular for good reason. They offer viewers a chance to relate to their favorite entertainers and to connect over shared laughter. In a world that is far too often far too serious, this kind of content serves as a light at the end of the tunnel.

The 24-Hour Challenges

As an entertainer, you might share a lot of your life with others online. However, the 24-hour challenge is designed to take this to the next level. Here, your goal is to take on unique marathon challenges that morph ordinary days into something a little — or a lot — crazier. Instead of going about your day like everyone else, you use the power of editing and creativity to create a spectacle that will have your audience craving more.

These challenges require you to document full days from sunrise to sunrise and the wild things that you can get up to within that time frame. You're going to invite your viewers to come with you, to follow you from start to finish while you do your best to keep your sanity. Collect highlights as you go to create a series of videos on your main channel, or livestream the whole experience so that the most dedicated viewers can follow along.

This gives them a chance to join you for a whole day, to give them an exclusive window into the weirder parts of your life. Whether you plan on taking your audience on a wild ride through 24-hour gaming sessions, or simply do a relaxed day-in-the-life, 24-hour challenges can take your content to new heights.

Take a look at the 10 suggestions below to find out what you can do to take this challenge to the next level — 24/7.

12. Spend 24 Hours in a Specific Location

Spend a day in a place that you've never been before, and bring your audience along for the ride by documenting it all. Maybe you have a favorite city to travel to, and you'd love to share every little hidden gem that you've found. Or you've heard of a haunted hotel in your area, and you enjoy the adrenaline rush of challenging yourself to spend a night in terror. There are hundreds of locations to choose from that could take your day from boring to incredible with a bit of clever editing and fun reactions.

Ask your audience to retell tales of experiences they've had while traveling, whether that's something supernatural, or a translation mishap that led to hilarious consequences. Encouraging viewers to share their own life experiences in the comments has the potential to foster a positive community with great engagement, which will increase not only your views, but secure a great reputation for your channel, as well.

13. The 24-Hour Food Challenge

You might be a foodie or a terribly picky eater, or somewhere in between. Wherever you are on the scale, embarking on a culinary adventure can make for an exciting watch for your audience. Try different foods and meals for 24 hours, documenting your reactions and offering reviews along the way. Once you've finished, curate a collection of the day's highlights and upload these highlights to your channel. You can even upload multiple parts, if needed.

You can also theme your day—whether that's trying fast foods from famous chains or diving into the strangest foods you can find. You can even go for something truly riveting and try all of the spiciest foods your city or town has to offer—but that's for those who are truly brave!

14. DIY Project Marathon

If there are DIY projects that you've always wanted to do, but keep putting off, this is the perfect opportunity to get going and create some fantastic content in the process. From repainting your living room to building an intricate shelf, there's a wide variety of options that will definitely take a while to complete. Document the whole process—all of the drama, the mistakes, and the triumphs—from start to finish.

Once you've completed your DIY project, collect the highlights from your marathon and curate a compilation that will keep your audience hooked. Edit it to sheer perfection and upload this collection of clips while encouraging viewers to comment on their favorite, or least favorite, part of the process. This will allow you to see which parts of your videos caught the most attention, and help you plan for the next project.

15. A 24-Hour Gaming Marathon

Gaming videos are a true staple of content creation on almost every platform. Get your energy drinks, a comfortable chair, and game face ready and throw on a 24-hour streaming session. You can upload highlights onto your channel afterward for an additional set of great content.

This offers you a chance to interact with your community and increase engagement through polls, asking for strategies and suggestions, or even asking your audience to throw you into chaos by voting for mods or choices that you need to make in-game. You can even add charity fundraisers to this marathon and encourage viewers to donate while they watch. The possibilities really are endless with this one!

16. A 24-Hour Fitness Challenge

If you're into health and fitness, or even if you're not, showcasing an entire day dedicated to it can really elevate your content. Try out new hiking trails, train for a marathon, attempt new workouts, and share tips for your audience to stay active. Make sure to document every step of your journey and pay attention to the effect of the weather on your audiovisual elements if you're recording outside. Upload the most important, insane, or beautiful moments as a set of highlights.

You can even add in elements of maintaining a healthy diet, adding fresh meal ideas and keeping up with hydration. Whether you're known for your fitness passion or are more of a couch potato, this kind of content is bound to draw an audience looking for inspiration to stay active.

17. The 24-Hour Travel Adventure

Think of a destination that you've always wanted to go to, and all of the ways that you could possibly get there. Perhaps you want to take a train ride across the country, or you want to try out flying to remote airports. Or you'd simply like to go on a road trip with friends or other creators. Make sure that you know the rules of recording in public in the country or area that you're visiting and stay courteous to others before setting up your camera.

Record aspects of your trip that might be challenging, exciting, or simply relaxing. Talk to your viewers about what you're experiencing and encourage them to share their own travel tales or nightmares in the comments. Ask for tips and tricks for new travelers, too, for any specific destination that you're going to or for traveling in general.

18. Try a 24-Hour Tech Experiment

Technology has absolutely exploded in the digital age. There are all kinds of gadgets, devices, and apps to try out, some for particularly unique purposes. You can find the strangest gadgets on Amazon or any other e-store and spend a day testing them out—or even use them exclusively *instead of* your normal go-to devices. Perhaps you even want to take a trip down memory lane and swap out smartphones and Wi-Fi for vintage tech from the '80s and '90s to showcase just how reliant we've become on our electronics.

Record yourself using these technologies and every moment of nostalgic joy or sheer, unadulterated frustration. Make sure that your recording remains high-quality, however, though you can add elements of other forms of recording that you discover. Perhaps add a clip of you using an old video phone or add images taken with ancient Polaroid cameras. Ask viewers to comment about the technologies that they remember from their childhood, or perhaps even strange tech they used to use.

19. A 24-Hour Art Marathon

If you're an entertainer, odds are that you have at least a bit of a creative streak in you. An art marathon is a fresh idea that will allow you to produce a range of new art pieces, crafts, or illustrations that can truly inspire and wow your audience. Record the process of you creating art, from the materials that you're using to the finished product, and beautifully document it all.

If you've never tried painting before, or have never done anything of the kind, you can show your viewers that everyone begins somewhere, that passion and dedication can make a real difference in your life. Create a mural in your home, or attempt to knit a scarf, or even try your hand at woodwork. Your finished creations can even serve as a source of new merch for your brand!

20. Play a 24-Hour Music Challenge

Have you ever wanted to try out an instrument, or are you a musician who's passionate about creating incredible sound? Try a 24-hour music challenge where you explore new genres, create your own compositions, or simply share your musical journey with your audience. Document this process by recording samples, doing a mock-show, or simply sharing your experience on camera.

You can showcase what it takes to learn a new instrument by completely dedicating your day to trying out beginner's courses and attempting to hold a tune. You can also ask for suggestions for new genres to try out and offer your own reactions and reviews to entertain your community. Music is a true source of human connection; don't be afraid to use it!

21. Host a 24-Hour Q&A Session

Host a livestream session for a full 24 hours wherein your audience can ask you anything and engage in conversation with you. This can help to build a strong and relatable community around your channel and humanize you for your viewers. It also creates conversation and can bring up new video and content ideas that you might not have thought of yourself. Beyond all of that, it offers a chance for serious introspection and to see the perspectives that other people have of you when they watch your content.

All of these challenges are marathon undertakings that give you the opportunity to create incredible long-form content. You can even divide these days into several half- or full-hour chunks to upload over a specific period of time. Having longer sessions like this gives your audience a chance to see more of who you are as a person and fosters an important sense of connection with them.

Food Challenges

Food challenges are an exciting, larger-than-life way to create a community of inspiration, fun, and sheer entertainment for your audience. Maybe you've never even cooked an egg and definitely will not be a 5-star chef; it doesn't matter. Food challenges are a great way to create content for *anyone*. Even if it isn't a niche that you usually create in, viewers are always interested in the culinary skills— or lack thereof — of their favorite video-makers, and that's something you should seriously take advantage of.

These challenges can be as creative and exciting as you want them to be. Create every foodie's dream, or become your very own kitchen nightmare — as long as it is *entertaining*, your audience will watch. The great thing about food challenges is that they can be as simple or as complex as you want them to be. You can try out the spiciest snacks you can find, or try to make a menu item from a gourmet restaurant. Or maybe you want to test your stomach by eating as many hot dogs as you can.

You can judge your own creations or bring friends and other entertainers together to critique your skills. Interesting your guests will drive more views, shares, and opinions from your audience. Ask them for recipes that they'd like you to try out or challenges that they think you should be taking on when it comes to food; the opportunities for engagement are almost endless.

This chapter will encourage you to tackle great feasts, brave the spiciest spices and to experiment with the weirdest and most bizarre flavors out there. Get your taste buds ready and try out these 8 challenges to satiate your viewers' appetites!

22. Spicy Food

Maybe you've spent your life trying to eat the spiciest foods you can, or perhaps the most flavor you've added to your food is a touch of ranch. Either way, testing your mettle with the hottest of the lot is sure to capture the attention of any audience. From hot wings to burning candies, the world is your oyster. You can even bring in a friend or two, or even your father-in-law or strangers on the street, and see who can really handle the heat.

Record yourself and other participants eating the spiciest foods you can find, or perhaps spicy food that has been suggested to you by viewers. Use the magic of editing and sound to increase the tension and explosive impact that the hot spices might have on you. Ask viewers to comment on the spiciest things that they've tried, or ask for suggestions for future videos.

23. Fast – Food Roulette

Gather up a group of other creators or your own friends and family for a fun and unique get-together. Spin a wheel—whether a digital one or one you physically created—to determine which fast-food restaurant you'll be trying out, or which menu items everyone has to eat. This will add a lovely element of surprise that will have viewers wanting to watch it all unfold. You can even buy a wide selection of fast food beforehand and play a variety of games to see who gets what!

Record the challenge itself, or sprinkle in the preparation, if you want to include that in your video as well. Collect the best moments into an exciting compilation to upload. Remember while recording to ask viewers to comment on their favorite fast foods and restaurants and place this request at strategic spots within the final video. This allows viewers to feel as if they're a part of the challenge, and it will improve engagement.

24. Giant vs. Miniature

Who doesn't love a bizarrely huge or an unusually small version of regular-sized things? Well, food is no exception. Record yourself searching and finding the biggest and smallest versions of food in your area. Perhaps your town is famous for its absurdly enormous burgers, or you know a good recipe for a very tiny pancake. Either way, hilarity will come with absurdity when you combine these elements with their normal-sized counterparts.

While recording—either in your home or in a restaurant—compare and contrast the taste and the experience of these different-sized foods, and determine once and for all whether bigger (or smaller) is truly better. Encourage audience members to comment on portions, and ask for suggestions on the next comparison that you can explore. This will improve engagement numbers *and* can possibly offer you new avenues for even more content!

25. One Color

It might seem easy to create a dish that has only one color, but it might be tougher than you think. Try out a five-course meal based on one single hue, and try to find *only* ingredients in that color to use. It might take some ingenuity to create something edible, never mind truly delicious. Record the adventure and share your thoughts on the process and the taste experience. You can also use this opportunity to add an educational twist to your content by informing your audience about fun facts regarding certain colors of food.

Your viewers will be highly entertained by watching the challenges you face along the way. Perhaps you can even host a party and encourage friends to bring along foods based on different colors of the rainbow and see which color truly stands out above the rest. Ask viewers to discuss their favorite foods and favorite colors, which could perhaps lead to a bit of good-natured banter in the comments. This is a unique way to foster community, connection, and engagement on your videos.

26. Foreign – Food Taste Test

Set up recording equipment around a table arranged with various candies, snacks, and foods from all over the globe. You can even theme various videos after various countries, and offer your honest reactions and reviews for unfamiliar flavors as you test each one.

You might even insert a call to action by asking your audience which country or region you should try out next, and encourage them to engage in your comments. This journey across the world offers viewers the chance to witness diverse cultures and dishes—and might even inspire them to try things out for themselves!

27. DIY Mystery

This is another enormously fun and entertaining group activity for content creators of all ages and backgrounds. Ask your friends or family to create mystery dishes with specific predetermined ingredients, and get together for dinner. Record both this process as well as the experience of the food itself. Compile the best moments for your audience to enjoy.

Before diving into your meals, everyone gets a chance to guess what others have made. For an added layer of intrigue, you can combine this video idea with the blindfold challenge. This will offer you a chance to showcase your creativity and to engage in a bit of friendly competition that's sure to titillate your audience.

28. No Utensils Challenge

If you're not afraid to get messy, this challenge can be truly entertaining and fun, both for your audience *and* for you. Find meals that would be particularly difficult to eat without any utensils at all and rely only on your hands. There are even ways that you could create added difficulties for yourself. Make sure to set up your recording to catch all of the hilarity up close and personal while you're participating in the challenge.

For example, you can add a timer to the challenge, to see whether you can finish every crumb or morsel before the clock ticks down to zero. Or you can remove your hands as well and see whether you can eat the whole meal using only your mouth! Either way, this challenge is sure to have your audience roaring with laughter as they watch your culinary struggles.

____

____

____

____

29. Extreme Food Combinations Challenge

Many people have strange food habits, combining odd things like ranch and fruit loops or seafood and chocolate. Ask your audience to submit strange flavors for you to try out, or scour the internet for the oddest culinary combos to try out. You might be surprised by which ones you find delicious, even if they are downright bizarre.

Choose the best or strangest options and set up a meal for yourself in front of your camera. Offer your reactions and reviews for pairings, and encourage your audience to engage with their opinions in the comment section.

Food challenges aren't just for extreme foodies; they can add an exciting and unique element to *any* channel. They give your audience a chance to get to know you in a new way and can bring new subscribers in through entertaining antics and fascinating content. Go out there and give it a go—you might thoroughly enjoy it!

Challenge Videos With a Twist

Adding a bit of a twist to challenges will make your content more unique among the many thousands of videos that are out there. Keeping your audience hooked is always necessary, and throwing a few curve balls at them is one of the best ways to do that. Challenge videos are a YouTube staple, but you should be bringing something fresh and exciting for your viewers to enjoy—something that defines your own style.

In this type of video, you're playing the ultimate expectation-versus-reality game, and the quality of your twists and turns will determine whether viewers stay or click off. As the creator of the video, you need to make sure that your audience feels the thrill of the emotional rollercoaster that your challenges promise.

Make the unexpected a reality, and invite your viewers to try and guess what will happen next. Keep them on the edges of their seats and take them on a wild ride through your imagination with these 8 challenges.

30. Escape Room Challenge at Home

Bring the drama and excitement of escape rooms right into your living room with puzzles and challenges that will blow your viewers' minds. Have friends or family join you, create teams, and compete against one another for the ultimate prize of being the escape room champion. Make sure to have recording equipment at strategic spots in the escape room—or multiple rooms, if you're creating an entire maze.

Introduce unexpected obstacles, like mysterious visitors or hidden rooms that block your progression, to create great twists and turns. This will surprise both the audience and the participants, skyrocketing the excitement and tension of your videos. Ask viewers to pause the video and submit their guesses about clues in the comments to gain high levels of engagement.

31. Cooking With Mystery Ingredients

Set up a dinner or a lunch with friends, family, or other creators where you'll each cook a dish using mystery ingredients. Let everyone choose between boxes labeled with question marks or other containers wherein the ingredients won't be visible. Then, participants can see what ingredients they have, without letting others see them. These ingredients must then be used to create something delectable for everyone to enjoy.

Halfway through the challenge, add a surprise ingredient for each participant. This way, even if things are going suspiciously well, you've thrown a bit of a wrench into the works and created extra drama for your audience. Once all of the cooking has been done, sit down to enjoy the meal, and vote on who did the best and worst jobs. Encourage viewers to voice their own opinions about this topic in the comment section.

32. Random Object Olympics

Turn everyday objects into sports equipment and host your own makeshift Olympics. Set up obstacles like a broomstick high jump or duct-tape limbo. Create teams with friends, family, or other entertainers, and get together to compete in an entertaining and highly silly competition.

Participants can even dress up in their own costumes to represent various teams or countries. When doing these brand-new sports, you can throw unexpected rule changes or obstacles into the mix to add a fun twist. This is bound to bring a spot of comedy into the challenge to keep your viewers watching and laughing. Ask viewers about their favorite competition or what they think you should include in the next random Olympics.

33. DIY Obstacle Course

Why not grab random items that you can find in or around your house and build your own obstacle course? You can make it even more exciting by using polls or live streams to have viewers choose your obstacles from Amazon, Walmart, or other stores. This way, you can create content from building the obstacle course, rather than just the final challenge.

Of course, you can throw in extra levels of drama and suspense by blindfolding the participants just before they're ready to go. This way, neither they nor the viewers know what's coming next!

34. Reverse Charades

Throw tradition out the window by twisting around the well-known game of charades. Instead of having one person act out a scenario while the others guess, have one person guess and the others *all* act out the word. This added level of chaos will increase confusion and hilarity and create incredibly entertaining frustration—especially if your viewers already know the word that's being acted out. You can do the same thing with games like 30 Seconds or Pictionary, taking game-night classics and giving them new life.

35. Two Truths and a Lie: Extreme Edition

Everyone knows this little ice breaker, but there are ways to make it even more interesting. You can do it yourself by hosting a live stream and offering your viewers two outrageous truths and one plausible lie. Then, they can guess which is which through comments or polls.

You can even make it into a game with other creators, friends, or family. Don't go for subtle differences; try to make the truths even more unbelievable than the lies. This lets your audience connect with you and get to know you on a deeper level while offering maximum entertainment at the same time.

36. Invisible Object

Just about everyone loves mimes, right? Well, why not turn this ancient art form into a challenge? Have participants act like they're holding or interacting with invisible objects. You can throw a bunch of words into a bowl and have participants pull one word each from these. Other participants then have to guess what the invisible object could be.

Add in a twist by introducing additional, unexpected scenarios that the participants have to put themselves in—while still interacting with their invisible object. For example, perhaps someone pulled the words "coconut tree," but you give them the scenario of trying to board a plane. Hilarity is guaranteed to ensue.

37. Puzzle Maze

Build your own maze and add DIY puzzles along the way for participants to figure out before they can move forward. To make things a bit more exciting, you can add time constraints to increase the pressure or tie pairs of participants together. Now, they have to compete together—but, alas, nobody is allowed to speak for the duration of the challenge. This will have your audience both highly engaged and frustrated as they watch the puzzles unfold in front of their eyes.

38. Shout Singing

This is a singing competition with a wild twist. Instead of actually *singing* the songs, shout the lyrics out instead. Have other participants join in by trying to guess which iconic or rare songs you're shouting. This is also an idea for a live stream where the audience has to guess the shouted songs from a few lines of lyrics. You can also have a version where one person is wearing noise-canceling headphones while the other tries to sing, and have them guess the song just from mouth movements.

These challenges add elements of surprise to delight and entertain your audience, keeping them engaged for hours and eager to see how participants will navigate unexpected twists. What are you waiting for? Get out there and create proverbial roller coaster rides for your audience to enjoy!

Costume Challenges

If you like dressing up, then costume challenges are the perfect niche for you. They're fabulous, fun ways to bring your imagination to life and show off your creativity. Whether you want to be dark and moody or throw glitter over everything, costume challenges are open to you. They let you be a fashion chameleon, turning yourself into anything from a villain to a backyard fairy and more.

You can keep it simple by offering your viewers advice for their daily style, helping others find an affordable way to express themselves. Or, you can go all-out and create something magnificent, over-the-top, or even downright bizarre in a unique video series that will keep 3399. your viewers wondering throughout.

Dress up as your favorite characters and invite your audience to vote on their favorites, or use ordinary objects to create something really fancy. Whatever you can think of, there's a costume challenge for it. But, of course, everyone has to start somewhere.

So, here are 16 entertaining costume challenge video themes for you to pick from.

39. Random Costume Generator Challenge

The internet is full of random generators, whether you're looking for names, cosplay ideas, or brand-new characters that no one has ever seen. Use this to your advantage in your next video by searching for the perfect randomizer and using it to select unexpected combinations of themes and characters.

Record this process before revealing to your audience the final costume that you will bring to life. Finally, create your incredible piece of art and try it on, showcasing your interpretation of the combination of elements on camera. Ask viewers to discuss their favorite and least-favorite aspects of the outfit in the comments section. You could even create community polls for your audience to vote on.

40. DIY Budget Costume Challenge

There's nothing that challenges your creativity in fashion quite like a truly tight budget. Set a particular amount for yourself to strive toward while crafting a unique outfit, and let your viewers know exactly what it is while you're recording. While creating this beautiful and affordable set of garments, document the cost of each item and calculate the total at the very end.

Try to keep within the bounds of your budget while pushing your creative energy to its very limits. Recycle and repurpose everyday items or dig through your closet for old, barely worn clothing to craft something new. Ask your viewers to comment on their best tips for saving money and still looking great.

41. Cosplay Transformation Challenge

Choose a beloved cosplay character, whether they resemble you or not, and document your step-by-step transformation into that character. While recording, share your tips on costume assembly, wig styling, makeup, and logistics, if the costume makes it particularly difficult to move about.

There are thousands of incredible and popular characters across a wide variety of genres that you can imitate through your own artistic expression. Because of this, a cosplay challenge might be the perfect opportunity to draw a wider audience to your content. Invite newcomers to share their favorite characters and shows in the comments to illustrate a welcoming community.

42. Historical Fashion Time Travel

Humans are fascinated by history, and it should be no wonder why. Choose a particular historical period that has always tickled your fancy and examine the fashion trends from the age. You can incorporate your research into your video by sharing fun facts and describing the reasons behind various fashion choices from that time. This will add a creative educational touch to your videos.

Once you're confident in your knowledge, showcase the process of creating a costume from scratch inspired by the trends of your chosen era. You can even create skits illustrating a typical day or interaction of the time, using clever editing to simulate the perfect background. Ask your viewers which era they would most like to have lived in and encourage discussions about it in the comments.

43. Celebrities in Everyday Situations

This is a great challenge to ask other entertainers or your friends to join in on. Have participants dress up as famous celebrities and record your portrayals of these individuals in humorous, everyday scenarios.

Host a dinner party, go out to lunch, or create a skit where you're all standing in line at a grocery store. This creates a lighthearted take on the lives of the rich and famous, sure to entertain and amuse your audience. It also allows your audience to imagine what their lives would be like or how it would change if they one day became truly famous—or infamous.

Encourage viewers to discuss what they think the lives of the rich and famous are really like behind the scenes. This will improve the video's engagement and draw more views to your channel.

44. Gender-Swap Character Challenge

Find an iconic character from a television show, a movie, a cartoon, or even a legendary commercial and switch it up by swapping their gender through costume. This will bring a fresh perspective to a well-known persona and offer you the opportunity to role-play and entertain your audience with something new and unique.

Challenge other creators to swap traditional gender roles and transform into characters they might typically associate with the opposite gender, breaking stereotypes and allowing for heightened self-expression. Ask your audience to leave their opinions in the comments and ask for more ideas on new characters and personas to explore.

All of this will bring more eyes to your channel, particularly if you choose to collaborate with others, and will open an important discussion in your comments. This leads to higher levels of engagement and a stronger place within the algorithm.

45. Literal Interpretation Costume Challenge

Find often-used phrases or idioms and use them as inspiration for a more literal approach to costume design. For example, think of "comparing apples and oranges," and have a friend dress up as an orange, while you portray an apple. This way, you can turn language into a visually entertaining experience for your audience.

You can add a bit of education into your recording by researching *where* various phrases and idioms came from, and *why* they became common to use. You can even try your hand at interpreting slang phrases in this way and look at how language is evolving in the modern world. Ask your audience to vote on the best interpretation of an idiom at the end of the video, encouraging discussion in your comments section.

46. Video Game Character Transformation

Video games continue to be a popular form of media. Choose the most well-known character you can find, or go back to the 1980s and find something obscure. Transform into these characters by dressing up in a self-made costume. This allows you to illustrate your skills in makeup, costume design, and prop creation.

Ask your audience to guess which character you're portraying—which can be a bigger challenge if you've chosen a rare, unknown character—and tell them to leave their guesses in the comments section. Reveal each character at the very end of their section, or even at the end of the video, to keep your audience guessing throughout. This will improve both engagement and retention, ensuring that viewers stay and watch rather than clicking away halfway through your videos.

47. Fusion Costume Mashups

Search for two or more completely unrelated themes or characters and document the process of crafting a hybrid costume. Blend the various elements in amusing and unexpected ways to create something unique and entertaining. Let your audience follow along as you face the challenges of trying to bring together such vastly different universes into a harmonious and beautiful expression of artistry.

You can make more videos like this and ask your audience for ideas on future combinations. You can even have other entertainers or friends record with you once you're fully in costume and ask them to guess the characters that you've incorporated into your outfit. This will draw their audiences to you and your audience to them, which will increase views and subscribers all around.

48. Decades Fashion Showdown

Fashion, like everything else, has evolved over time. From the Roaring '20s to disco in the '70s and even butterfly clips in the '00s, all kinds of trends have popped up over the decades. Travel through time by exploring and showcasing these iconic fashions by creating a runway in your living room. Record yourself showing off various outfits from each era and add little fun facts to the screen through editing to add a bit of education into the mix.

To connect with viewers through fashion nostalgia, ask your audience about their favorite looks from the past and things that they wish never actually existed. Share your own opinions on looks from the various decades and point out your favorites—as well as those that you think are better left behind.

49. Historical Figure Dress-Up

This is another way to dive into history, but this challenge focuses on famous historical figures and their impact on the world. This unique combination of humor and education will offer fresh perspectives on the past and how it has shaped our modern world.

Find a figure from history whom you look up to, or who had some importance, and create an accurate costume to wear while you tell their life story, relay fun facts about them, or simply create a discussion about their role in the world. Encourage viewers to share their opinions about this person or other figures in the comments section to drive engagement and discussion on the video.

50. Cartoon Character Makeovers

Record the process of creating a costume of a beloved cartoon character from your childhood. Show off the final product in a re-enactment of one of your favorite scenes—or perhaps a mock interview between you as yourself and you as the character. You can be truly creative with the materials for the costume and the makeup that you use.

Ask your audience to tell you about their childhood favorites in the comments to create nostalgia through engagement. Looking back fondly at the past allows for a culture of mutual interest to form within your community, which will strengthen it.

51. Fantasy Character Fashion Show

Escaping from reality through fantasy is one way that many of us deal with the negativity and struggles that we face. Using this escapism to inspire and awe those who view you can really elevate your content to a higher plane.

Embrace the realm of the fantastical through documenting the process of designing and creating costumes that have been inspired by incredible worlds, fairy tales, and mythical creatures. These costumes do not have to be accurate to an existing character; you can use them as a form of true creative expression and bring your own interpretation to the table. Ask viewers to give their opinion on your various pieces and to suggest a myth or fantasy tale that you should base the next costume on.

52. DIY Superhero Showdown

Bring together other entertainers and friends and unleash your inner superheroes. Film yourselves competing to imagine the best original superhero characters, and then create their costumes using everyday items. Each superhero must have their own unique powers and backstories.

You can show off the finished costumes through humorous skits, having each participant play out a scenario in which their powers would be most useful. Clip the final costumes and the best moments from the video into a TikTok video, Instagram reel, or YouTube short, and create polls for your community to vote on whom they think should win the competition. This will increase your engagement across all platforms and strengthen your community.

53. Animal Kingdom Dress-Up

Explore the animal kingdom by dressing up as a variety of animals for a wild twist to the traditional costume challenge. You can choose popular animals according to a particular theme—for instance, African animals, big cats, or sea creatures—or you can search for the planet's most unusual animals to bring a unique element to the video, like the blobfish.

Add in fun facts about each animal that your audience might not be aware of to sprinkle in a bit of education to the entertainment. Encourage your audience to share weird and wild animal facts in the comments, and to discuss which animal you've portrayed should be ranked as the best.

54. Unexpected Theme Mash-Ups

Challenge the status quo by combining the most unexpected themes together to create quirky and hilarious costume combinations. For instance, perhaps you'd like to dress as a pirate who loves Hello Kitty or a whale in the desert. This will offer you a chance to really push your creativity to its limits and give your viewers a unique take on more traditional genres.

Request more strange mash-up ideas from your audience in the comments, and encourage them to think of the wildest combinations in the world. This slight element of competition will drive more engagement and build excitement for your viewers.

All of these costume challenge themes have been created to offer you a delightful mix of creativity, humor, and entertainment. They're meticulously curated for viewers who appreciate the art of dressing up, and for entertainers who want to channel their inner fashionista. The sky's the limit, so, get out there and get designing!

Extreme Weather Challenges

Some entertainers prefer creating their videos in the comfort of their own homes. However, there are others who like something a little more... adventurous. Extreme weather challenges are the epitome of adventure for those who seek an adrenaline rush unlike anything else.

If you're able to do it safely, your audience will be unable to look away from their screens while you brave the craziest elements that Mother Nature has to offer. Whether that's enduring heat waves, surviving freezing temperatures, or getting up close and personal with a tornado, these challenges are guaranteed to leave your audience craving more after every video.

These videos allow you to bring your viewers out of their rooms and take them on an adventure that they can only imagine—from the comfort of their couches, beds, or wherever they might be watching from. Audiences enjoy danger—if they're not the ones facing it. If you're brave enough, extreme weather challenges can help you build a community that will follow you everywhere—virtually, of course.

Below are 9 exciting themes for your next extreme weather challenge video.

55. Arctic Survival Challenge

If you live in a northern country, this challenge is possible to do at home during the winter, but you can take it a step further by traveling all the way to the Arctic. Record yourself as you brave freezing temperatures and tough conditions.

Test your survival skills in the snow and ice, showcasing your strategies for staying warm and safe. Make sure that you do your research beforehand and that you always have emergency measures in place, especially if you want to travel to more remote areas that can get far below freezing.

Share knowledge on survival in these conditions and discuss your experiences, the challenges you're facing, and things you would change if you were to try it again. Ask your audience what they'd like to see next and whether they'd ever attempt an Arctic challenge.

____

____

____

____

56. Desert Endurance Challenge

If you live close to a desert, this challenge is fairly easy to do. If you have the means, you can also travel to one of the most famous deserts in the world, like the Sahara, the Gobi, the Sonoran desert, or even the Namib Desert. Document yourself as you endure the incredible heat and face challenges related to dehydration, heat exhaustion, and navigating a harsh landscape. In some deserts, this will include climbing sand dunes, while in others, it will mean walking across very hard ground.

Before you attempt this challenge, ensure that there are safety measures in place and that you research survival techniques and strategies. Collect enough equipment, water, and food for the time that you will be spending in any remote region.

Share your experiences, difficulties, and the strategies that you've learned with your viewers. You can also tell them about the desert that you find yourself in by sharing fun facts in a humorous way. Ask viewers whether they would want to live in a desert and if they think they'd be able to survive.

57. Tropical Storm Shelter Build

Document the process of collecting materials and equipment, and then build a sturdy shelter that is capable of withstanding tropical storms. Throughout the recording, discuss the necessary building skills you need to build this shelter and how you can effectively prepare for storms. Share which materials are more suitable and why, and discuss fun facts about tropical storms.

Ask your viewers to share their experiences with storms in the comments, and encourage them to stay safe and make sure that they do thorough research before attempting to build their own shelter. This will show your concern for your audience's safety and a sense of empathy that will build stronger connections with viewers.

58. Blizzard Camping Adventure

This challenge requires quite a bit of preparation and research; it offers a great opportunity for a multipart series culminating in an exhilarating grand finale. Record the process of preparing and then head off to camp in the middle of a snowstorm, highlighting the challenges of setting up your camp, staying warm, and cooking in extreme winter conditions.

Offer advice to viewers regarding what to do if they ever find themselves stuck in a similar situation. Discuss the best strategies for retaining heat and the kind of food you should take with you if you're going camping during the winter. Remind viewers to stay safe and not to attempt anything without professionals present. Make sure that you have your own emergency measures in place throughout the process of recording.

Ask viewers whether they'd prefer to live in freezing weather or scorching weather. Encourage discussions about the weather in various regions and countries in the comments, too.

59. Thunderstorm Photography Challenge

In this challenge, your aim is to capture the intense beauty of thunderstorms through photography, combining the thrill of extreme weather with the artistry of visual storytelling. Take your audience with you on the journey to capture the perfect picture, explaining every step in the process as you go.

Share the challenges, failures, and successes you experience throughout, letting viewers see a glimpse into your mind as an artist. Add the final results of your photography journey into your video, or create a slideshow for YouTube Shorts, TikTok, or Instagram that you can link back to your YouTube channel. This will enable you to draw audiences from various platforms and to build a cross-media community.

60. Lightning Storm Nighttime Exploration

Journey through the outdoors during a night-time lightning storm while documenting the experience. Ensure that you have emergency and safety measures in place to limit any possible dangers. Showcase the awe-inspiring visuals of lightning to your audience, while educating them about staying safe during these storms. Explain the physics of lightning and the reasons behind the incredible show that it can create in the sky.

Encourage viewers to share their own lightning videos and photos on their social media with a particular hashtag, and ask them to describe their experiences in the comments. This will draw more views to your videos and increase engagement across your platforms.

61. Hurricane Preparedness Drill

Gather equipment and supplies and run a realistic hurricane preparedness drill. Show your audience everything they will need to be safe during a hurricane, and explain the steps to securing property, gathering emergency supplies, and planning evacuations. Discuss what you think the most important priorities should be in the event of a hurricane and how to do everything safely when a storm is already raging.

You can also include facts about hurricanes or tales of historical hurricanes in your video for an additional twist and higher entertainment or educational value. Ask viewers about the craziest hurricane facts they know and emphasize the importance of being aware and educated about extreme weather.

62. Extreme Wind Resistance Test

Document yourself as you test the wind resistance of various structures, objects, or personal gear in high-wind conditions, demonstrating the power and impact of this form of extreme weather. Make sure that your safety measures are *always* in place and that you inform your viewers of the safety strategies required in rough wind conditions.

Use the opportunity to educate your audience about various forms of equipment and windy conditions, and offer tips and tricks for when they might be stuck in a precarious situation. Ask viewers to share their experiences with particularly gusty conditions and what they did to remain safe. You can also shorten the information you've given into bullet-point lists to share on YouTube shorts, Instagram, TikTok, or your other social media.

63. Tornado Alley Expedition

Pack up your gear and film your journey as you head off to Tornado Alley to showcase the unpredictable nature of tornadoes. Explore the science behind these powerful storms while heading on a road trip through the area. Share how this form of travel can be done safely and show your audience all of the measures that you take to ensure that you aren't whisked away by the winds.

Discuss the differences between real tornadoes and those you see in movies, as well as the differences between tornadoes and other storms or forms of wind. This kind of education can be clipped into smaller chunks as short fun-fact videos for other platforms, too. This way, you can increase the reach of your content and create a larger community.

Extreme weather challenge videos not only provide an adrenaline rush but also offer valuable insights into survival skills, preparedness, and the impact of different weather conditions. They offer audiences something exhilarating to watch while they are safe in their own homes. These are experiences not many people have in their lifetime, but they're something that you as an entertainer (if you do it safely) can harness as an expression of the strength of the human spirit.

Whisper Challenges

If you've ever played a game of telephone, you probably know how funny miscommunication can be. Whisper challenges take this misheard silliness to a new level. On the one hand, you have someone wearing the strongest noise-canceling headphones known to man. On the other, you have someone trying to convey a message to them using only whispered words.

The misinterpretations that can come as a result of this are guaranteed to be entertaining, hilarious, and sometimes downright weird. Honestly, it's a goldmine of content just waiting for you to start digging. The frustration that you'll experience will all be worth it when you see those analytics skyrocket.

Audiences will love watching the struggle to understand, especially if you choose a particularly difficult challenge. This genre's unique angle also means that you can really play around to find your own style and even twist more classic challenges.

So, get your noise-canceling headphones ready and prepare yourself for what lies ahead with the 10 ideas I've laid out below.

64. Movie Quotes

Here, you and a friend set yourself across from each other. One person will wear the notorious noise-canceling headphones while the other whispers the world's most iconic movie quotes to them. The listener must guess which movie the quotes are from. These quotes are guaranteed to get lost in translation, with the listener doing their best to decipher words from just the movement of the whisperer's mouth.

Ask the audience to play along with the guessing and to indicate how many they got right in the comments. Encourage them to discuss their own favorite movie quotes and to offer more ideas for your next whisper challenge.

65. Song Lyrics

With this challenge, you can bring other entertainers, friends, or family into the mix and separate them into teams of two. Contestants have to attempt to whisper popular song lyrics to their teammate, who is wearing headphones. That teammate must then guess the song before the other team can. Dramatic renditions should be awarded bonus points. Whoever has the most points at the end of the game wins!

If you bring other entertainers into the competition and live stream it, you can draw more views and use highlights to boost the analytics of your channel. Other video makers participating in the competition can then upload their versions of the competition as well. This way, you'll be able to grow your collective community.

66. Foreign Languages

This one is a bit more of an *extreme* challenge. Participants have to whisper words or phrases in a foreign language, and the teammate wearing headphones must not only guess *which* language it is but also attempt to translate it into English. For higher entertainment value, you can add matching body language to help the guesser out.

If nobody speaks a foreign language, there are language apps that can assist you—even Google Translate, though that might not be entirely accurate. Ask your audience to talk about where they're from and which language or dialect they speak in the comments. You can even ask them to share stories of misunderstandings stemming from multilingualism. This will increase global engagement and encourage viewers from other countries to join your community.

67. Tongue Twisters

Gather a group of friends or other video-makers together and split up into teams of two. Test participants' enunciation skills by having them whisper tricky tongue twisters to their teammates. These teammates, armed with their noise-canceling headphones, must then attempt to decipher the tongue twister correctly.

The team with the most points at the end of the challenge wins! You can improve engagement and connection with your community—especially the international community— by asking viewers to comment on their favorite tongue twisters either in English or in their own language. You can also add an educational element by discussing the origins of the various tongue twisters used in the challenge.

68. Celebrity Impressions

In this challenge, participants have to whisper catchphrases from famous celebrities or impersonate their favorite celebrities, while other participants must then guess who that celebrity is. You can divide into teams or have one friend join you for a co-op challenge. Adding body language will sprinkle in a humorous twist to an already funny bit of content.

You can make the challenge more difficult by choosing niche or micro-celebrities, or even your favorite fellow YouTubers. Throw in a call to action by asking the audience to pause the video after an impersonation and add their own guesses in the comments. You can also create polls for viewers to vote on their favorite impressions in the community section.

69. Sound Effects

Choose a theme, like action movies or animal noises, and then gather a group of friends to participate. Some participants have to wear headphones, while others attempt to re-create sound effects from the theme using only whispers. Those wearing the headphones then need to guess what the intended sound was. Cue lots of amusing attempts at whispering barks, explosions, and perhaps even the ocean's tides.

Add in a twist by filling sound effects onto pieces of paper and throwing them into a bowl for participants to pull from when it's their turn to make the sound. This can add a new level of hilarity and frustration into the mix. You can also encourage the audience to guess the sounds themselves and let you know in the comments section how many they answered correctly.

70. Mystery Object Description

In this challenge, two participants sit across from each other. One participant must think of a mystery object and describe it to their teammate using whispers. The teammate must guess what the object is based only on the clues whispered to them. You can make things even more thrilling by having the whisperer pull the mystery object's name from a hat and giving them a time limit to describe said object. If they guess correctly, they will win points; if not, they lose them. Whoever has the most points at the end of the game wins the challenge.

Encourage your audience to talk about the easiest and most difficult object to guess in the comments section. You can even create polls with this question in the community section and on your other social media to increase engagement and to pull viewers from other platforms back to your YouTube channel.

71. Historical Events

Search for a series of historical events or facts related to a particular theme—for instance, wild weather or political figures. Then, gather a group of participants and split them into teams. For an added element, teams can dress up or give themselves a name. One team member wears headphones while the other whispers these historical events or facts.

The listener must then guess the event, time period, or fact being described. You can make it as easy or as difficult as you'd like! These facts can then be separated into clips or bulleted lists for more short-form content on other platforms. You can encourage engagement from your audience by asking them the weirdest fact they know about history, or what their favorite time period is and why.

72. Jokes and Puns

In this challenge, you have to gather the best—or the cringiest—jokes that you can find, whether from people you know in real life or by scouring the internet. Print these jokes out on sheets of paper and gather a group of participants. Each team should receive one of the sheets of jokes.

One teammate then has to whisper them to the other, while the other person has to attempt not to burst into laughter while guessing the punchline. You can also do this without teams by simply collaborating with another video-maker and whispering the jokes to them. Include a call to action by asking viewers for their cringiest or best jokes in the comments section. You can also ask about their favorite jokes from the video, and which one was the hardest to guess.

73. Meme Phrases

Take advantage of viral internet content by making it part of your next whisper challenge video. Collect a series of fantastic memes and viral catchphrases and get together with friends or other video-makers for this challenge. Each team gets a set of catchphrases. One person wears headphones while the other whispers the memes. The teammate with the headphones must then guess which meme it's from.

This can lead to all kinds of hilarity and perhaps even become a new meme in itself! Let your audience guess along by pausing the video and discussing the catchphrases in the comments. You can also use snippets from the video for your other social media to draw in more viewers.

These whisper challenge themes add a delightful twist to the traditional game, ensuring a mix of humor, confusion, and entertainment for both participants and viewers. You can use these fun misunderstandings to create more content and grow your community.

Reverse Challenges

The reverse challenge makes every day an opposite day. Flip the script on the usual by turning it all around through the magic of editing and rewinding anything that you can think of. Maybe you want to un-eat a meal, do your daily schedule backward, or even reverse the effects of a balloon popping in slow motion. In fact, reverse slow motion is a fantastic way to keep your audience interested.

The uncanny impact of reversing normal, everyday stuff is what makes these challenges so intriguing for audiences. Hit rewind on anything from making breakfast cereal to completing an obstacle course.

You don't need to completely rely on editing to get things done, either; some of these challenges offer you the opportunity to try your hand at reverse acting. Maybe you can have your friends try to say phrases in reverse, or even walk backward for a day. You can even recycle some older content by simply turning it all around.

What are you waiting for? Let's throw this world into reverse gear with the 9 ideas below.

74. Backward-Speaking Challenge

In this challenge, participants have to try to speak sentences or phrases backward accurately while viewers enjoy the hilariously garbled results. You can prerecord sentences and phrases and play them backward through the magic of editing, comparing them to real-life attempts.

Ask your viewers to discuss in the comments whether they think participants did a good job or not, and why. You can add in a sprinkle of education by explaining why it's difficult for our brains to understand or create backward communication, too. Or, you can level up the humor by using slow-motion effects!

75. Reverse-Eating Challenge

You can do this challenge by yourself or bring in a few friends. Create meals or snacks in "reverse style." For instance, you can build a sandwich with the bread on the inside and tomato, bacon, and lettuce on the outside—turning that BLT into a TLB—or perhaps a cookie with the icing on the top and bottom. Cue messy and amusing eating attempts that will surely have your audience roaring with laughter.

You can turn this into a series by asking the audience to suggest future reverse foods for your next videos and increase engagement at the same time. You can also ask them to name their favorite reverse food in the comments.

76. Rewind Dance Challenge

Find the most trending music videos or TikTok dances out there. Then, participants have to attempt to do these dances themselves. You can also choreograph an entirely new dance for them to try. You can document the entire process of learning the dance, too, if you'd like.

Once the footage has been recorded, use editing software to reverse the final dance to showcase the comical and awkward dance moves that are created when they're played backward to the same song. Remember to make sure about the chosen music's copyright—using copyright-free songs might be a better choice.

You can ask your audience whether they enjoyed the usual dance or the reverse dance more, and to talk about their favorite moves in the comments. Who knows—your hilarious creation might become the next infamous TikTok dance!

77. Reverse-Drawing Challenge

Channel your inner SpongeBob SquarePants by attempting to draw a picture or create an artwork in reverse. Begin with its final details and reverse the process, finishing with the initial outlines. You can also film the process of creating art entirely, and then use editing to play this process in reverse.

If you choose this form of the challenge, you can also explain the steps of the process in reverse, starting with explaining how you created the final product and ending with how you began its planning. Encourage your audience to discuss their favorite and least-favorite parts of the artistic process in the comments.

78. Backward Clothing Challenge

Change the face of fashion by either wearing your clothes backward or by putting on your clothes in reverse order, beginning with your shoes. With this challenge, you can expect a lot of laughter and fashion faux pas. You can also have a fashion show, or have a competition to see who wore it better—for the latter, you can create voting polls for your community across your social media accounts to generate more engagement.

Perhaps you can even attempt to do everyday tasks wearing your backward wardrobe, like cooking, walking around your house, or even doing a yoga routine in the living room.

79. Backward Obstacle Course

Create an obstacle course in your backyard and have a few participants navigate through it in reverse—from the finish line to the starting line. To make it even more challenging, ask participants to do the whole thing while moving backward, facing the finish line.

This will create hilarious attempts at overcoming the obstacles in a unique way and will have both you and your audience roaring with laughter. Add in a commentary character like a real sports event for even more entertainment value. Ask your audience which obstacle they think was the most difficult to navigate backward, and whether they'd want to take part in a challenge like this.

80. Reverse Makeup

In this challenge, you have to attempt to apply your makeup in reverse order. This means that you'll start with lipstick and end with your foundation. At the same time, you need to attempt to make it look as good as you can.

Experiment with different makeup looks in this challenge and ask your audience to vote on which one looks the best. Offer tips about various makeup steps while you're recording the challenge, explaining why it would usually be done in a certain order and why it might not work that well in reverse.

81. Reverse Storytelling

Find a popular short story, children's story, or even a Reddit post that you particularly enjoyed reading. Read the story to your audience in reverse order, beginning with the last sentence and working your way backward.

Not only will this create a unique narrative experience, it might give the story's twists and turns an entirely different feeling. You can create a series from this idea by requesting suggestions for future stories to read backward from your audience in the comments.

82. Backward Sports

Gather a group of your friends or collaborate with other video-makers in this challenge, in which you attempt to play sports in reverse. Whether it's playing soccer backward, swinging a golf club the wrong way around, or shooting a basketball with your back facing the net, the results are sure to be amusing and unexpected.

You can also use editing to reverse trick shots that participants perform and slow it all down to show the amazing mechanics of sports. Ask your audience about their favorite sports and what else they'd like to see you play. If you're collaborating with other video makers, encourage your audience to visit their channels, and they will do the same for you.

These reverse challenge themes are sure to add a playful touch to creative ideas, offering your viewers a uniquely entertaining experience. They also open a world of new reverse opportunities for your channel from which you can create collaboration with others or even build your own set of series. Go out and show them how it's done—backward!

Art Challenges

Art challenges are a chance for you to discover your creativity and to show the vastness of your imagination to your audience. Using your artistic tools as your weapons, you can push the boundaries of reality. These challenges give you the power of inspiration and allow you to dive headfirst into creating masterpieces.

From exploring techniques and themes to experimenting with mediums and materials, art challenges are limitless opportunities. Re-create famous artworks in your own style, attempt a bit of speed-drawing, or teach yourself new ways to express your thoughts. Whatever you can think of can be turned into a beautiful challenge that your viewers will love to watch.

Invite your audience to join you as you push your limits and unleash your imagination with these 9 art challenge ideas.

83. Three-Marker Challenge

Gather together all the markers that you have available and throw them into a bowl or a box that you can't see into. Randomly select three markers from these and challenge yourself to create a vibrant and cohesive artwork with only those colors. You can do something similar with colored pencils, crayons, sidewalk chalk, or even paint tubes. Whichever medium you choose, make sure that you use only three colors to create the final product.

You can even attempt to create this artwork in different mediums, each time only using three colors. Ask your audience to discuss their favorite piece or color. You can add tips and tricks for making art in different mediums or with particular colors for an educational twist.

84. Speed-Drawing Challenge

Witness the artistic process in fast motion by challenging yourself to complete a highly detailed drawing within a limited time frame. You can choose one specific medium, or try to see how fast you can do similar drawings through different media. For instance, you can compare your speed on a drawing tablet to when you're physically drawing on a piece of paper or painting on canvas.

Throughout the process, discuss the complexities that go into creating art, the challenges that you face having to work at such a quick speed, and the reasons why some media can take longer than others. Ask your viewers to comment on ideas for future speed-drawing challenges in terms of what they'd like you to draw, the medium they want you to use, or even the time limit that you need to set for yourself.

85. Palette-Knife Painting

Explore the unique and textured results of trading a traditional brush for a palette knife. Create beautiful paintings with an unconventional tool, discussing the differences that you notice with your audience. You can also add an additional twist by using a variety of different tools to create different pieces of art. From creating dot pieces with a marshmallow to creating textures with faux fur, there are endless possibilities out there.

Share the challenges and triumphs of using unconventional equipment and ask your audience which strange tool they think created the best picture. This can lead you into a fresh, new world of exploration and expression that will give you a fresh perspective on art creation.

86. Tiny-Art Challenge

Bigger isn't always better. Challenge yourself to push the limits by creating miniature versions of your usual artwork. Try to shrink it down as much as you can, strategically planning the equipment you'll use and even investing in a magnifying glass, if necessary. Showcase your precision, talent, and attention to detail on a small scale.

Use the opportunity to share the difficulties of creating such tiny pieces as well as how the experience has affected you as an artist. Ask your audience to take part in the discussion in the comments by speaking about the value of art and whether size should impact that. This will lead to thoughtful discussion while remaining engaging and interesting.

87. One-Color Art Challenge

Limit your color palette to only one color and create visually appealing and even diverse artwork with it. This is a way to test yourself and push your creativity while simultaneously offering interesting content for your audience to consume. Explain the use of color theory and how different textures and brushstrokes can affect the final result.

You can turn this into a series by choosing a theme for each video and asking your audience to contribute their ideas for theme and color combinations in the comments. This will help you connect to your viewers and bring in fresh ideas and new perspectives.

88. Collaborative Art Challenge

Reach out to other video-makers or gather a group of artsy friends and join forces to create a collaborative piece. One person begins the artwork, while the others contribute in turns to demonstrate the beauty of artistic teamwork. You can do this in person by using a canvas, or during an online stream by creating digital art. You can even create multiple pieces by each starting your own, then switching the pieces between one another every ten minutes or so.

With multiple pieces, you can create voting polls on your social media, or ask your audience to talk about their favorite part of each piece in the comments. You can also ask for suggestions for more future collaborations in terms of artists they'd like to see join or in terms of themes for the artwork itself.

89. Art-Swap Challenge

Reach out to another artist to collaborate with you. Exchange your usual art supplies or styles with them and showcase the unique fusion of techniques and perspectives to your audience. For example, if you're a digital realism artist, find an artist who prefers painting fantasy on canvas, and switch those around.

Document your experiences and discuss what it's like to work in this new style and on a different medium. Showcase appreciation for various forms of art and the talent and dedication needed to create something amazing. Ask your viewers about the styles and mediums they enjoy, whether they're artists themselves or not.

____

____

____

____

90. Replicating Famous Art

Find the most iconic pieces from art history and re-create them in your own style and on a medium that you're most comfortable with. Share interesting facts while documenting the process of re-creation, telling your viewers about the history behind the piece and the reason for its fame. You can also create a series from this, by creating videos on the most famous artists from different time periods and their impact on the art of their time.

This is a great opportunity for both personal artistic exploration *and* adding a sprinkle of education to your audience's entertainment. Ask them what the weirdest fact is that they know about famous artists or which artist's work speaks to them the most.

91. Inktober Challenges

Utilize the viral nature of monthly prompts or challenges like the annual Inktober challenge, where artists create daily ink drawings based on given themes throughout the month. These challenges reach a much wider community and can bring more viewers to your channel specifically. It also gives you an opportunity to network with other video-makers in your space and to experiment with various themes and ideas.

Encourage your viewers to give their input during these events to create fresh ideas for the next theme. You can even create limited-edition merch for audiences to purchase based on each event that you participate in.

Art challenges offer a mix of creativity, skill demonstration, and entertainment, both for the video-makers who participate and for their audiences. They offer an escape from creative ruts and a way for you to explore your own expression by traveling down new and unique avenues. So, choose your challenge and let the creative juices flow!

Blindfolded Challenges

Blindfolded challenges take away your sight, but they also allow you to create a new kind of vision. It's a wild ride from start to finish with something unexpected waiting around every corner. In these challenges, participants are blindfolded from the start and bravely need to stumble through the tasks they're given. From trying to guess different types of foods to attempting a blind obstacle course, every challenge is designed to build suspense and reveal surprises.

Your audience has the chance to come along and watch as you and other participants struggle through everything without the ability to see. It's all about uncertainty and the difficulty of doing things without one of your important senses.

The 9 themes below have been carefully curated to create hilarious and binge-worthy watching for your audience.

92. Blindfolded Makeup

Gather a group of friends or collaborate with other video-makers, or even attempt this challenge on your own for extra difficulty. Challenge participants to apply makeup either on themselves or on someone else while blindfolded. This activity is one where the visual aspect is truly important, so the results are guaranteed to be surprising and hilarious.

Once everyone has completed their looks, reveal the final product and point out where things might have gone wrong. Discuss the difficulties you faced and the frustration of not being able to see what you were doing. Encourage viewers to talk about their favorite looks and if they think you've done a good job.

93. Blindfolded Taste Test

Gather a group together and collect a series of foods or meals from restaurants. Keep the food out of view until everyone has been blindfolded. Then, present the food and have participants taste a bite of each meal. They have to guess what they're eating, and, if you want to level up the difficulty, you can even have them guess which restaurant it's from.

Use camera angles and close-ups to catch the best moments and to let your audience experience all of the reactions as participants have to navigate through various different flavors. Encourage your audience to point out the best moments—you can use those mentioned most often as clips for YouTube Shorts or other short-form video platforms.

94. Blindfolded Obstacle Course

Build an obstacle course and have participants navigate their way through it while blindfolded—just make sure that you have the proper safety measures in place! You can add an additional element by having a friend navigate them through the obstacle course by shouting commands or code words from the finish line.

Expect plenty of stumbling, laughter, and unexpected detours on the journey through the course. Encourage viewers to comment on their favorite obstacle and to suggest more obstacles for the next challenge. If you're collaborating with other video-makers, you can even invite your audience to subscribe to them, and vice versa, to build a bigger and stronger community.

95. Blindfolded Puzzle

You can do this challenge on your own or collaborate with others. In this challenge, you have to attempt to solve various puzzles or simply build a jigsaw puzzle while blindfolded. You can add in another element by asking a friend to guide you, but they are not allowed to touch the puzzle and have to use a predetermined set of commands to assist.

Ask viewers which of their friends they think would be the best communicator in a challenge like this and encourage them to talk about it in the comments, or to tell stories of situations where they had to rely "blindly" on their friends. This will foster connection, build a relatable image, and encourage engagement on your channel.

96. Blindfolded DIY

On your own or with friends, attempt to complete a simple (and safe!) DIY project or craft while blindfolded. Creativity will fly out of the window, leaving only unpredictability, hilarious frustration, and a bit of a mess behind. One friend might be reading or giving the instructions to those who are blindfolded, but the seeing friend is not allowed to physically assist in any way, or to deviate from the instructions if you're putting something together.

For instance, they can't suggest that you have to move your hand to the left to find the scissors, but they can say that you have to stick a circular piece of paper onto wood. Throughout the process, or after the final product has been revealed, share your frustrations and ask your audience to talk about the moments that they found the most dramatic or frustrating in the comments.

97. Blindfolded Guess the Object

Gather a few household objects or ask your audience for suggestions on odd objects to buy from Amazon or other online retailers. Then, bring together a group of friends, other video-makers, or even your parents and announce the competition. They have to feel the various objects while they're blindfolded and try to guess what they are, which can lead to amusing guesses and reactions. Whoever has the most guesses right at the end of the game wins!

98. Blindfolded Musical Chairs

Spice up the classic game of musical chairs by adding blindfolds into the mix. Now, participants have to navigate the chairs and find a seat without the sense of sight. The other rules remain the same, though you might need to put some safety measures in place to avoid any injuries.

Ask your viewers about other childhood games that you can add to your list of blindfold challenges, like hide-and-seek or Marco Polo, for example. You can create multipart videos from those suggestions and even add highlights to more short-form platforms to increase the reach of your content.

99. Blindfolded LEGO Building

You can either buy a LEGO set that's made to build a specific structure, or simply use basic LEGO bricks to build any structure you'd like—like a space ship or even a simple bridge. The catch is that you'll be blindfolded the entire time. You can also collaborate with others and put together teams to build various LEGO sets.

Hilarity will ensue while everyone tries to figure out the pieces and attempts to assemble them without seeing. You can also have a team member who reads the instructions if you are using specific LEGO sets. Encourage your audience to discuss the sets they remember from their childhood, like their very first LEGO set, or simply something that makes them feel nostalgic.

100. Blindfolded Art

In this challenge, you are blindfolded while you attempt to craft something amazing, like a sculpture, a painting, or even a crochet piece. You have to rely solely on your sense of touch and spatial awareness to create an artistic masterpiece while being unable to determine which colors you're using.

In this challenge, you have to pay attention to textures, movement, and feeling rather than sight, highlighting the importance of other senses when creating art. Encourage your audience to talk about this idea in the comments and share their own experiences with limited senses.

These blindfolded challenge themes offer a blend of humor, unpredictability, and creativity, making them entertaining for you and your audience. There are many forms of collaboration among them, which will better connect you with your peers and create a stronger network of video-makers working together, too.

No-Technology Challenges

The no-technology challenge is a bit like a digital detox. The point is to step out of the digital world and test your dependence on gadgets and devices. With this, you can show your audience how much technology has changed our lives by getting rid of it for a day—or perhaps even more.

There are no notifications, screens, or constant scrolling in these challenges. Instead, you have to focus on real life, whether that's in your interactions or in performing your day-to-day activities. Whether it's sending handwritten letters, communicating face-to-face, or navigating with a paper map rather than a GPS, you'll travel back in time to a place without buttons and screens for everything you do.

Viewers will love watching because it's both intriguing and funny. There are struggles and triumphs and a sense of slowing down and enjoying the world around you. So, put down your phone, tablet, smart fridge, and all of the other gadgets you might own, and try out these 10 inspirational ideas.

101. No-Technology Day

Spend a whole day without using any technology (aside from setting up your recording equipment, of course). This includes smartphones, computers, and any other electronic devices like your toaster, coffee maker, and even light switches, for 24 hours.

Document the challenges you face trying to navigate your daily tasks without using modern tech to do so. Explain the difficulties and the triumphs of cooking, exercising, and even trying to relax without turning to a screen. Ask viewers whether they'd try a tech-free day and which devices they think they spend the most time on.

102. Analog Gaming

Rather than setting up for a night of online gaming, or even solo gaming on your PC or console, gather a group of friends or family to participate in a traditional game night. Bring together old board games, card games, or puzzles, and spend an evening, or even an entire day, solely relying on these for entertainment—without ever touching a digital device.

You don't have to go for mainstream games, either; you can find very niche or rare games that not many people have seen before for an additional unique twist. Ask your viewers to tell you about their favorite games and puzzles from their youth to improve engagement.

103. Outdoor Adventure

Come together with a few others and embark on an outdoor adventure or a camping trip without any technology—except for recording equipment and one phone to be used in an emergency situation. Record all of your struggles, from setting up camp without digital instructions to exploring nature.

This will help you relax while showcasing the challenges and enjoyment of a tech-free weekend and an escape from the trappings of the modern world. Ask your viewers to share their dream destination for their own tech-free trips, or to share which one digital device they'd always want to have while camping, aside from their phones.

104. Cooking Without Technology

For this challenge, you need to prepare a meal without any modern kitchen gadgets, electronic appliances, or digital recipes. It might not seem that difficult, but imagine baking bread without an oven and an electronic mixer, or cooking up a full breakfast without a blender or a toaster.

You can share fun facts about historical cooking and how people circumvented the challenges of a kitchen lacking electricity for an educational twist. You can even create a series by asking your audience for suggestions regarding more historically significant meals that you can make without modern equipment.

105. No Social Media

In this challenge, you have to abstain from using social media platforms for a designated period of time, whether that's a day, a week, a month, or more—aside from uploading videos documenting the experience. Explore how your daily routine, social interactions, and overall well-being are affected.

Ask viewers whether they've ever taken a vacation from social media and how the experience impacted their lives. Of course, only come back to read and respond to comments or to discuss the experience once the designated period of time has lapsed—after all, YouTube is social media, too!

106. Digital Detox Challenge

For this challenge, you need to take a break from all of your digital devices, including your computer and smartphone, only setting up equipment to document the experience. Focus instead on offline activities like reading, drawing, or participating in other hobbies. Examine the impact that this has on you during the day and discuss the implications for the world at large.

Encourage your audience to take part in the discussion in the comments and to take their own digital detox break to see how that affects them, even if it's only for a day or two.

107. No Screens Before Bed

For a set amount of time, or perhaps even in an attempt to build a new permanent habit, set a time before bed when you will avoid using screens. This includes phones, tablets, electronic photographs, computers, the television, and even your smart fridge—if you have one.

Document the experience by evaluating the effects that it has on your sleep quality and relaxation. Share research and studies on the subject and their findings and compare it to what you notice in yourself. Encourage your viewers to try out this challenge and to share their own experiences in the comments section.

108. Tech-Free Creative Challenge

Engage in a creative endeavor like drawing, crafting, or writing without using any digital tools at all. Whether it's trying to paint or writing a 20-page short story, share your experiences and the difficulties of being without any of the usual tools that you might use for creation.

Even Googling inspiration pictures or the spelling of a certain word is off the table—meaning you only have your hands and creative prowess at your side. Encourage viewers to discuss their tech-free hobbies and to reflect on how often they use digital tools or electronic equipment when they're not even thinking about it.

109. No GPS Challenge

Find a new place to check out, like a park, a restaurant, or even a small vintage shop in your city or another city entirely. Document the process as you navigate to your destination without using a GPS or any map apps, revealing the grand adventure of getting lost and trying to find your way using only a paper map and your intuition.

Share your frustration in a comedic spectacle as you try to find your way while resisting the urge to glance at Google Maps. Ask your audience how they would navigate with only basic directions and access to paper maps.

110. Communication Without Technology Challenge

Most of us spend a lot of time texting, calling, or connecting with others through social media. What would happen if you spent a day without any of these technological advances and tried to communicate with your friends or family? Try to do so in this challenge and see how successful you are.

You can also try to send a few old-fashioned letters and track their progress as they head for their destination. Compare how long it takes the letter to reach the person you want to contact to how quickly we can contact one another through technology. Ask your audience to comment when the last time they sent a letter was—or if they ever even have sent one.

These themes for your no-technology challenge will encourage you and other participants to step away from the digital world, fostering creativity, mindfulness, and a deeper connection with the offline environment and will remind your viewers that life is multifaceted.

Virtual Challenges

On the opposite end of the spectrum from the no-technology challenge are virtual challenges. Here, you explore virtual places and things, showcasing the power of the human imagination to your viewers. Essentially, it's an opportunity with very few limits and a lot of new ideas to discover.

You can even combine virtual reality with the real world to put a brand-new twist on these challenges. Try everything from virtual escape rooms to VR tournaments to walking through the park with AR glasses—and take your audience along for all of it. Wherever your imagination can take you, virtual worlds make it possible.

Get ready to pixelate your reality with these 10 engaging virtual challenge themes.

111. Virtual Escape Room Challenge

In this challenge, you and your friends or other video-makers have to engage in an online escape room experience. You have to solve puzzles and virtually work together to complete the room within a set time frame. If you want to add a bit of competition, you can even split participants into different teams and try to beat the room before the other team does.

Ask your viewers to comment on how long it took them to figure out the various clues and which moments of the challenge were the most tense. You can also ask them to submit suggestions for more virtual escape rooms to try out.

112. Virtual Scavenger Hunt

For this challenge, you need to gather a group of people and set up a virtual scavenger hunt with cryptic clues. Give participants the first clue that will have them search for specific items in their houses, showing you what they find on a video call. Once they find the first item, they can get the next clue, and so on. Whoever finishes the scavenger hunt first wins.

Encourage viewers to talk about scavenger hunt experiences they've had in the past, or whether your clues were too difficult or too easy. This will increase engagement and help you gauge the sweet spot for your next challenges.

113. Virtual Talent Show

Host a virtual talent show in which participants show off their talents through virtual performances, whether that's singing and dancing or doing incredible magic tricks. Livestream the actual talent show and compile a set of highlight clips for your YouTube channel.

Ask your viewers to comment on who they think will win and ask them if they were right when you reveal the winner at the very end. This will keep viewers watching to see what the final results are by keeping them curious throughout the video.

114. Online Fitness Challenge

Host a virtual fitness challenge where you and other participants all follow an online workout routine together. You can take it easy or push the limits with a particularly difficult fitness challenge. You can also try out virtual sports games that encourage exercise while playing them. Throughout the challenge, you can ask participants to offer their best fitness and health tips, or encourage hilarity and make jokes if you are all mostly couch potatoes.

Ask your viewers which virtual workout they think you should try out next, or encourage them to discuss their own health tips in the comments.

115. Virtual Gaming Tournament

Organize an online gaming tournament in which participants will compete in various video games and stream the gameplay for their audiences to enjoy. You can bring together other video-makers who focus their content on video games, or even video-makers who only occasionally play casually. This can bring an element of humor and uniqueness to your content.

This also allows a cross-channel community to form, allowing you to draw viewers from other channels, and for those channels to draw viewers from you. This can create a new avenue for collaborative content creation, as well.

116. Virtual Cooking Class

In this challenge, you'll bring together other video-makers in a virtual cooking class where everyone needs to follow the same recipe. Participants cook together by video call and show off their culinary prowess in the process. You can choose well-known cooking channels to collaborate with or entertainers who are good sports and like to try out new challenges.

This will allow you to explore new connections with other video-makers while expressing yourself through the culinary arts. Viewers will also have the opportunity to engage with new content and other video-makers, both from your channel and from theirs.

117. Virtual Trivia Night

Host a virtual trivia competition wherein participants have to answer questions on various topics. You can have several trivia rounds with various formats and themes to really test their general knowledge and to foster friendly competition. You can also add in an element of fun facts to increase the education value, or make it particularly difficult for hilarious guesses and attempted answers to come into play.

Ask your audience what their favorite question was and how many of the answers they knew themselves. If you're livestreaming the event, have the chat answer the questions in tandem with participants. Clip a few facts and questions for more short-form content on other platforms. All of this will allow you to create more content from one filming session.

118. Virtual Language Learning

Most people are probably familiar with the memes about the Duolingo owl mascot. Use this virality to your advantage by embarking on a language-learning journey. You can take part in lessons on Duolingo or any other language-learning app and even compare apps with one another. Create multiple episodes where you illustrate the improvement of your language skills through virtual lessons.

Connect with other video-makers who speak the language you're learning and attempt to have conversations with them in that language as part of your efforts. This will gain you a wider audience while helping you improve your language skills. Ask your viewers which other languages they're interested in learning to foster engagement in the comments.

119. Online DIY Craft-Off

Host a virtual DIY crafting challenge where participants have to create unique crafts using materials that they have at home. You can set a time limit and live stream the entire creative process, with participants showcasing their skills through video calls. From this, you can create polls during the stream where your audience can vote on who they think has made the best progress and which of the projects they think looks the best so far.

At the very end of the process, everyone should show off their final products and explain why they made certain design choices.

120. Virtual Book Club

Get together with other video-makers, such as book reviewers or writers, and host a virtual book club with several episodes. During each episode, discuss a chapter of a chosen book, or even read parts of it together and give your opinions on it. This can foster a sense of community and intellectual engagement.

Ask your audience to give their opinions about the same book in the comments section, and encourage them to engage with the opinions given by participants in the video. This will improve engagement and build a sense of connection between you and your audience.

These virtual challenge themes offer a mix of interactivity, entertainment, and community engagement, making them ideal for video-makers who want truly engaging and collaborative content. If you want to build a network with other entertainers, then these video ideas are perfect for you. Get out there and work together!

Genre Fusion Showdown Challenges

In these challenges, your aim is to combine two seemingly unrelated genres into a creative masterpiece. Maybe you want to try out making a music video from a cooking show, or have a dance-off between two historical figures, or even create a sci-fi comedy skit.

Essentially, the secret is to perfectly connect elements from very different genres to create a piece of content that is unique, innovative, and, ultimately, highly enjoyable for your audience. You need to find the perfect harmony between wildly different ideas and show off your skills to keep your viewers hooked.

Let the showdown begin with the 10 themes below!

121. Sci-Fi Comedy Extravaganza

In this challenge, your aim is to merge elements of science fiction with comedy to create a hilarious storyline that's based in a futuristic or space-themed environment. You can build a one-person show using clever props and editing, or bring together other participants to create a unique and hilarious show.

Ask your audience whether they'd like to see another episode with a similar premise to gauge interest in the storyline. This will tell you whether your comedic chops in combination with the vastness of space will serve to keep the attention of your community, or whether you need to try again.

122. Romantic Horror Twist

These definitely are two genres that most people wouldn't picture together, and that's exactly what makes it a unique and creative combination. Create a suspenseful and thrilling love story that's filled with unexpected scares and twists, keeping your audience on their toes throughout.

You can even end the video with a cliffhanger and ask your audience to comment what they think will happen next, and create a series that will take their breath away. Take inspiration from your comments, but add your own voice and style to bring the unexpected to life.

123. Historical Fantasy Adventure

In this challenge, your goal is to transport your audience to a historical setting infused with fantasy elements. Create a tale in which you introduce magical creatures and quests within a specific historical context. You can research myths and fairytales from various periods and places in history to add education and cultural appreciation to your story.

Ask your viewers which other historical periods and places they'd like to see in your next videos, and create a series where you discover and explore the fantasy of history.

124. Mystery Musical Mashup

Channel your inner Sherlock Holmes *and* your inner musical maestro in this challenge. Create a story in which you've combined the intrigue of a mystery with the joy and emotional intensity of various musical numbers. This will lead to a unique storyline where characters express themselves through song in the middle of an important and fascinating investigation.

Encourage your audience to guess the end of the investigation in the comments and to tell you what their favorite song from the story is. You can even lace some of the musical numbers with investigative foreshadowing, encouraging viewers to go back and rewatch after the final reveal has been made.

125. Superhero Cooking Challenge

Whether by yourself or in collaboration with others, merge the superhero and cooking genres through creating characters with unique or outrageous superpowers who have to participate in a cooking show. You can even add in a culinary showdown for a fun and quirky twist, where characters need to overcome challenges associated with their superpowers to create the perfect dish.

For instance, perhaps the character with super strength struggles with the more delicate aspects of being a chef, or the character with laser eyes tends to burn their steak. This will create an entertaining and hilarious bit of unique content that viewers are sure to thoroughly enjoy.

126. Dystopian Romantic Comedy

Create a dystopian world with a bit more of a lighthearted touch. Film a story where, despite the insane circumstances of the world—or perhaps because of them—a pair of characters humorously finds their way to a post-apocalyptic romance for the ages. This opens avenues for bizarre and funny situations wherein the human spirit will still triumph through love, even if it seems a bit cheesy.

Ask your viewers to comment about their ideal post-apocalyptic date, or encourage them to discuss what's next for the characters in your story. This can lead to more content being generated from this premise, opening the possibility of a series.

____

____

____

____

127. Action-Packed Fantasy Cooking Show

Create a full cooking show or a set of cooking skits that are set in a fantasy world. Here, participants—chefs—will battle not only in the kitchen, but also in epic fantasy quests to gather unique ingredients. Give ingredients new names based in fantasy and offer the participants the list of quests that they need to complete for all of the ingredients to be gathered.

Encourage engagement from your audience by asking which recipe they'd like to see "fantasy-fied" next and what kinds of quests they think the chefs should go on next. You can even base the fantasy world in iconic fantasy settings like *Lord of the Rings* or even *How to Train Your Dragon*.

128. Film Noir Sci-Fi Detective Story

In this narrative, you create a tale in which you blend the classic Film Noir detective genre with futuristic sci-fi elements to create something both compelling and visually striking. With Film Noir traditionally being set in the past, you can offer a unique spin by adding bizarre sci-fi twists—and even time travel.

Bring your audience with you on the journey by leaving clues for them to decipher in the comments or to spot in the background that might lead to them solving more than just a single investigation.

129. Animated Mystery Adventure

Combine the comedy and styling of animation with the mystery genre by taking viewers on an animated adventure filled with enigmatic puzzles and plot twists. You can take the angle of classics like *Scooby Doo*, or give it a darker Tim Burton element with muted colors and sharp character features.

This will keep your audience on the edge of their seats as they try to solve the mystery along with your originally created characters, which you can even incorporate into your next set of merch, if they're popular enough.

130. Thrilling Comedy Heist

Write a high-stakes heist storyline with an infusion of comedic elements, where characters have to navigate through humor-filled challenges while executing a daring robbery. You can even create a thoughtful satire piece by referring to huge channels like Troom Troom, who have infamously created a life hack video based on kidnapping.

Ask your audience to participate by voting for the next great heist that your new characters can go on, or ask for suggestions for another nefarious scheme.

These genre fusion showdown ideas offer you a delightful mix of various storytelling elements that will give your audience something fresh and new to enjoy. Each one has the potential to become a series of videos and to reinvigorate the image of your merch. So, why not grab an idea and go wild with it?

Entertainment and Reactions

Reaction videos put you in virtual front-row seats. You, as the video-maker, are suddenly the audience, sharing your reactions in real-time with your viewers. The internet is vast, and you can find the strangest and funniest things on earth in every nook and cranny. React to viral trends, movie trailers, music videos, or just people with odd lives, and offer your audience your own authenticity as they watch with you.

These videos connect you to viewers in a unique way, letting them feel like friends who enjoy watching content with you rather than strangers. So, give them your unfiltered commentary and bring your audience along as you create a universal watch party with the 6 ideas below.

131. Reacting to Old Photos

In this time-traveling escapade, dive into either your own past or that of a celebrity or even another creator. Show your audience the dusty relics of the past by unearthing old photographs from yesteryear. Rediscover old hairstyles and fashion choices and offer your unfiltered opinion on how they've aged.

Ask your audience about their most embarrassing or memorable moments from the past, and invite nostalgia into your comments section. This allows you to build a relatable image and to connect with your community through shared experiences.

132. Parodies

Parodies are comedic chameleons when it comes to content creation, where creators disguise themselves as satirical characters and change the serious into the absurd. Search for the most bizarre or funniest parodies that you can find from across the internet and invite your viewers to experience them with you.

You can also ask for suggestions on strange or hilarious parodies that your audience thinks you should react to next and take this on board for more content creation in the future.

133. Nostalgia Trips

This is a way for both you and your audience to get a bit sentimental about the past through modern rose-tinted glasses. Find old forms of technology, ancient movies, commercials from the '90s, or any other invention from your youth—or even your parents' youth. Invite your viewers to explore and experience these vintage finds with you.

Compare and contrast the way content was created back then with current trends on the internet. Comment on the tricks that were used to make certain scenes and moments feel more real for audiences. Ask your viewers to tell you about their favorite older movies or television shows and to describe them in your comments section.

134. Guess the [Blank]

Present your viewers with cryptic riddles and strange scenarios that will both amuse them and keep them on the edge of their seats. Whether it's predicting the outcomes of particular videos, identifying mystery objects in photos, or completing sentences or dialogue in movies with unexpected twists, you can be a true puppeteer of anticipation.

You can react to old music videos and have your audience guess lyrics, or even ask them to guess the ending of rare and niche movies and shows.

This will increase engagement and encourage your audience to come back for more unique content.

135. Music Covers

Find popular covers of songs on YouTube or TikTok, or even re-creations or reworked versions of those songs. You can also go for covers that have fewer views, but be careful not to punch down if you choose that route. Give your opinions and constructive criticisms of these covers, and ask your audience for their input as well.

If you find amazing covers, don't be afraid to share them and to give their creators more exposure and a bigger audience. This kind of wholesome action can come back to you in the long run, too, if they send viewers to you as well.

136. Life Commentary

In this kind of video, you set up your comedic courtroom, where you are the judge and your audience is the jury. Find videos of people behaving strangely, like rude customers, odd encounters, or even police body cam footage, and discuss the events with your audience.

Encourage viewers to leave their opinions in the comments as they watch the spectacle develop alongside you. Watch wild events unfold and don't hold back on your authentic reactions.

Reaction videos are popular because they create a sense of friendship, community, and relatability between creators and audiences. They cover anything from hilarious moments to serious topics and can open important—or simply funny—discussions between people in the comments. Don't be shy; tell the people what you think!

Part Two

YouTube Ideas for the Creatives

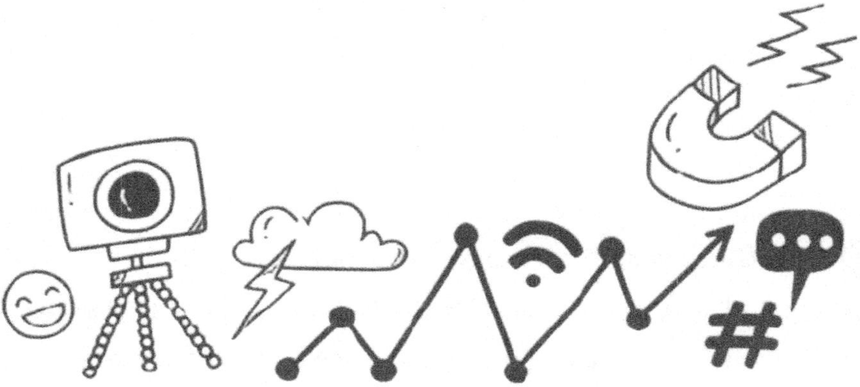

DIY Hacks

DIY hacks are a fantastic way to give your audience both content *and* great practical advice through creative solutions. This kind of content can keep viewers watching for a long time, especially if you consistently offer ways for them to build new skills that are useful in their daily lives. As you dive into these video ideas, arm yourself with a glue gun, a hammer, and some nails, and use your imagination to solve problems that everyone faces.

There are solutions for everything, from wobbling tables to easier ways to clean messes. Maybe you've got a unique way to make affordable smartphone stands out of everyday items, or you're a master of repurposing anything around the house—whatever it is, it's always a great idea to share your knowledge.

Be clever and innovative, and invite your viewers along for the journey with these 10 ideas.

137. Productivity Life Hacks

Film yourself sharing a set of hacks and tips to boost productivity in daily life, using practical examples to illustrate your points. Show your audience how to organize their workspace, create efficient schedules, and use productivity tools that they can use on their digital devices.

Use real-life examples by taking your audience on a journey through your day, showcasing what works best for you and offering various alternative options. Ask your audience to share their most helpful productivity tips and tricks that might help other viewers in the comments. This will create an atmosphere of connection and relatability.

138. Travel and Packing Tips

If you're a fan of travel, then this option could be great for you. Whether it's local destinations or international flights, many people don't often think of *all* of the challenges that you might face while traveling. Create a video in which you research and think of innovative solutions, whether you've used them yourself or not, and share the results with your audience.

These tips and tricks can include ideas for packing efficiently, choosing and organizing travel essentials in the best way possible, and making the most of limited space during trips. Ask your audience which tips they found the most useful and also what type of travel tips they'd like to see from you next.

139. Tech and Gadgets Hacks

In this challenge, your goal is to showcase clever ways that viewers can enhance the functionality of their gadgets, solve common tech issues, or even repurpose their old devices for new uses. You can choose everyday devices for a relatable perspective, or even find strange gadgets for a bit of an unusual twist.

Ask your audience about their favorite devices and the hacks that they think would be useful for their everyday lives. You can even ask them to tell you about the tech issues that plague them the most, and use that for your next video.

140. Kitchen and Cooking Hacks

Create a set of practical tips and hacks to save time when cooking, to better organize your kitchen, and to make the cooking process more efficient and enjoyable. You can make it even more useful for your audience by showcasing budget-friendly options and easy, healthy meals that can come together easily make their days less stressful.

Ask your audience about their favorite easy meals and the things they do to make being in the kitchen a bit simpler. This offers you a chance to connect and foster community.

141. DIY Home Decor and Organization

This is a very popular type of video, and for good reason. There are endless possibilities in the world of home decor. Whatever your aesthetic, you can use your knowledge of repurposing furniture, must-have organization solutions, and DIY decor to help your audience transform their living spaces.

Create a video where you illustrate your best tips for making a house into a home, and ask your audience what they'd like to see next in the home decor niche. Encourage viewers to personalize their homes and to embrace their quirks.

142. Fashion and Clothing Hacks

For this idea, you need to film yourself performing clever clothing hacks, offering styling tips for different people and creating DIY alterations that can refresh a wardrobe without breaking the bank. Many audiences will appreciate a budget-friendly way to give themselves a new look, and there's also the option for creating a series with this content.

Engage with your audience about different types of styles that they'd like to see but that are usually too expensive for the average person to dive into.

143. Beauty and Skincare Hacks

For this one, you can gather the most popular skincare products and use them, offering reviews for how they worked for you. While doing this, offer your audience beauty tips, options for skincare routines, and DIY beauty hacks with common household items.

Encourage your audience to talk about what works for them so that your community can help one another, something that will foster a positive reputation for your channel. Ask what they do to achieve healthy, radiant skin and what other health-hack themes they'd like to see from you.

144. DIY Photography and Videography Tricks

As a content creator, you likely have a large store of creative photography and videography hacks at your disposal. Use this knowledge by creating a video in which you explore these hacks. You can also offer up tips and tricks for low-budget equipment solutions, lighting tricks, and editing techniques that can help your audience.

Encourage viewers to make their own content shine and to reach for their dreams. Create a positive atmosphere for your viewers to learn from you as well as from one another.

145. Car and Automotive Hacks

If you enjoy talking about cars, create a video where you offer your viewers practical tips for maintaining their car, organizing it, and customizing it. Showcase the tricks you use to optimize your vehicle for convenience, efficiency, and aesthetics. Clip various tips and tricks for more short-form content, and encourage your audience to share their own vehicle tricks across the rest of your social media platforms.

You can also offer tips and tricks for things related to vehicles—for example, efficient fuel use, optimizing space in a garage or a parking lot, or even tips for better and easier driving.

146. Fitness and Wellness Hacks

For some, it's difficult to embark on a fitness and wellness journey, and many people have no idea where to start. You can help your audience begin by creating a video where you share innovative fitness hacks, workout routines, and wellness tips that they can easily incorporate into their daily routines.

Encourage your audience to share their progress and to talk about what has helped them on the road to wellness—this will help you foster a positive community. You can also share your own journey and showcase what has helped you the most.

These DIY hack video themes cater to various interests and needs, and they will have your viewers looking to you for practical solutions. If your suggestions are proven to work well, this will expand your reach and your subscriber base. So, don't delay. Grab your tools and get going!

How-To Guides

How-to guides have helped millions of people learn new skills, from how to do the dishes to how to fix any PC problem to how to plant their own indoor garden. These guides are a way for content creators to turn complicated ideas into easily digestible videos that viewers can follow to solve a myriad of problems.

As a creator, you can become a digital guide by sharing your knowledge in any subject that you're an expert on. Maybe you're fantastic at origami, or you fix up computers as a hobby; or maybe you know the steps to creating the perfect Christmas tree. Whatever you're brilliant at, you can turn that into helpful, fun-to-watch content for your audience.

Guides are also a great way to retain attention if you make content for people at different skill levels, because that will keep people watching for weeks, months, or even years to come. So, why not test out your teaching skills with the 9 ideas below?

147. DIY Home Improvement

Create a guide in which you take your audience through the steps for a variety of home-improvement projects. This can include anything from painting and furniture building to installing fixtures and upgrading decor on a budget. It's a great opportunity for a long-running series, as there is such a wide range of home improvement and maintenance tutorials for you to create.

You can even ask viewers to throw their hats in the ring by suggesting or asking for tutorials on specific projects that they might want to take on. This improves retention and encourages viewers to come back every time.

148. Technology Tutorials

Everyone uses technology, but not everyone knows the ins and outs of *how* to use it effectively, or how to fix it when it goes wrong. If you're a tech wiz, you can use this to your advantage by creating a video where you offer step-by-step instructions on using and troubleshooting various gadgets, software, apps, or devices.

Make sure that your tutorials cater to beginners, intermediate users, and even tech enthusiasts, ensuring that your content retains viewers as they continue to learn. Encourage your audience to let you know about common issues that they face, and use the most-mentioned ones as ideas for your next videos.

149. Cooking Techniques and Recipes

Cooking is something almost everyone has to do, but not something everyone knows *how* to do. And many of us often run out of ideas for new meals or simply hate all of the effort it takes to get things done in the kitchen.

You can help by creating instructional cooking videos that cover specific ideas, various recipes, and tips and tricks that can help your audience enhance their culinary skills in the kitchen. Ask your audience to request new meals for you to make in the comments, or to request tips on specific things.

150. Crafting and DIY Art Projects

When it comes to arts and crafts, the only limit is your imagination. Because of this, there is an incredibly vast number of projects that you could turn into videos. Choose a particular craft project and lead your viewers through every step from the beginning to the end.

Explain the materials and tools that they will need and explain every technique used in detail. Encourage viewers to request projects for the next video and ask them to comment on whether anything was difficult to understand or accomplish. This will help you refine and improve your instructions for the next one.

159. Language Learning Guides

If you're bi- or multilingual and have a passion for teaching, language-learning guides can be a great avenue for content. Offer basic, intermediate, and advanced language-learning tutorials, including guides on grammar, pronunciation, and vocabulary. Creating different levels of difficulty will keep viewers watching as they work on mastering a new language.

You can add elements of entertainment and fun activities into your videos to make them more exciting and unique. Cover both general conversation topics as well as more nuanced conversations to truly widen your audience's linguistic abilities.

160. Gardening and Plant Care

For those creators with a green thumb, teaching others how to care for plants is a great way to make engaging and beautiful content. Offer your viewers comprehensive guides on gardening, caring for specific plants, and creating the perfect landscape. Weave in tips and tricks on selecting, planting, and maintaining various types of plants, as well as how to create any aesthetic using particular sets of plants.

Ask your audience about the type of garden they would most like to have, and which plants struggle most with caring for in their gardens. This will improve engagement as well as offer you further themes for your content.

161. Beauty and Makeup Techniques

Beauty is a particularly popular niche in the YouTube space; it offers a wonderful creative outlet for your personal expression of style. If you're a beauty wiz, it's always a great idea to share your knowledge with others.

Create detailed makeup tutorials, beauty routines, and skincare guides for your audience to follow. In each video, make sure that you address different skill levels and cater to diverse beauty preferences, so that you can build a strong community of viewers from all walks of life.

162. Fitness and Exercise Routines

Many people want workouts and fitness routines that they can easily do at home, since not everyone enjoys going to the gym, and some people simply don't have the time. Take this opportunity to create workout guides, exercise routines, and fitness tutorials for viewers who want to try out new workout styles, achieve specific fitness goals or who simply need guidance on staying active.

Encourage your viewers to try out a new routine from you a few times a week, or do a few minutes of a shorter video every day. This will increase retention and encourage your audience to continue coming back for more.

163. DIY Fashion and Sewing Projects

In these guides, you will showcase how your audience can create their own fashion pieces, alter their own clothing, or try out sewing techniques. Illustrate every step of the process of creating something new from basic materials, offering tips and tricks to make things easier along the way.

Ask your audience to submit their designs through your other social media or to request specific pieces in your comments. Use this as inspiration for your imagination to go wild, and illustrate how you would interpret those designs or pieces. This will help your audience better connect with you and build a stronger community through fostering togetherness and creativity.

These how-to guides cover a wide array of interests, providing viewers with practical knowledge and skills to enhance various aspects of their lives. It's the kind of content that will pique interest and bring audiences back for more as they continue to build their passion and their skill. So, find what you're great at, and share your knowledge!

DIY Crafts

DIY crafts are a way for you, as a creator, to truly express yourself by sharing your masterpieces as content. You can use expensive crafting supplies or anything lying around your home to show off your creativity and use your imagination to keep viewers hooked.

There are so many options when it comes to crafting, too—from upcycling to fiber arts to building book nooks and carving wood, the potential for content creation is almost endless. On top of that, it will always be unique since every piece that you create will become a symbol of your own personal style.

If you have mad DIY skills, then this kind of content is perfect for you. Bring your audience along as you show your process every step of the way, and share the finished product across as many platforms as you want. Eight brilliantly creative ideas are described below.

164. Paper Craft Projects

Explore the world of paper crafting by creating tutorials on origami, paper cutting, and creating intricate paper-based designs. Showcase that paper crafting isn't as simple as your audience might think. Film the process of creating complex art pieces from the beginning to the end, including the materials and equipment that you use.

Offer tips and techniques to your audience throughout the process, so that they can make their own paper art. Show off the final products that you've created with great camera angles, and even build your own background. Ask your audience what type of project they would like to see next.

165. Upcycling and Repurposing

Most people have old pieces of furniture or objects lying around their house, or have access to a thrift store with interesting pieces at reasonable prices. Use this to your advantage by creating a video where you find a piece in your home or an antique store, perhaps even an auction. Then, illustrate the process of upcycling and repurposing this item to turn it into something new, functional, and beautiful.

Encourage your audience to try it out for themselves, illustrating the benefits, both creatively and financially, of upcycling old furniture. Offer tips and techniques that can help make the process easier and more fun.

166. DIY Home Decor Crafts

Find a set of materials from a craft store, or even things that are in your home already. Film this collecting process and discuss how these items can be used to create handmade home decor items such as personalized ornaments, wall art, or even decorative vases.

Showcase your creativity by bringing your viewers along for the process of creating fresh decor and how you've chosen to display it in your home. Ask your viewers which aesthetic they prefer and what kind of decor they'd like to see you create next.

167. DIY Gifts and Greeting Cards

This idea is particularly good to use during holiday seasons, though you can also create birthday-themed videos surrounding it. Gather materials and inspire your viewers with your ideas for crafting personalized and thoughtful gifts and handmade greeting cards for special occasions.

You can theme each video according to a particular holiday and discuss how you'd go about deciding on a type of gift according to various personality types, hobbies, and interests for particular people. Showcase creative ways to wrap gifts and how to add a personal flair to each one.

168. Polymer Clay Creations

Polymer clay accessories and creations are currently quite popular. Capitalize on this by sharing your process as you sculpt unique jewelry, figurines, and decorative items using clay. Showcase how you gather materials, the kind of paint and tools you use, how you shape the clay and let it set, and how you make sure that the item will last a long time.

If you have a small business related to your creations, showcase the craft fairs you attend and the ways in which you gain more customers. Encourage your viewers to visit or buy from you if they see something that they like.

169. Crochet and Knitting Projects

This is another area of crafting that has gained a lot of traction in the past few years. Take advantage of this popularity by catering to yarn enthusiasts. Create guides or tutorials on crocheting and knitting projects from scarves to plush toys. Ensure that these creations are unique and that you're not stealing content from other fiber artists.

Unique tutorials and discussions on yarn will ensure that your audience comes back to see every new item that you've created so that they can follow along and create these items for themselves. You can also use this to funnel more viewers to your other social media, or even to your own website.

170. DIY Jewelry-Making

Jewelry is one of those crafts that can be made with almost any material. From paper bracelets to pearl earrings to pasta necklaces, almost anything can become a great accessory. Try your hand at this craft by choosing various materials and creating jewelry from them.

Compare the styles, material types, and aesthetics that you use and review each one. Search for techniques and tips and record yourself trying them out, offering your audience an opinion on what you discover. Ask viewers to suggest more material types and accessories that they think you should make, which will foster connection and engagement.

171. Seasonal and Holiday Crafts

Take advantage of various seasons by trying out craft ideas according to specific seasonal and holiday themes. Try out DIY festive decorations, gifts, and other projects like blankets or furniture. Search for inspiration and ideas across the internet and showcase yourself attempting to re-create them.

Offer your own opinion and reviews of these ideas and ask your audience for their input on which ideas they think they'll use for their DIY holidays. You can also create your own crafts and offer short tutorials or guides on how your audience can create these on their own.

These craft ideas cater to a diverse audience, including those looking for inspiration and those who are seasonal crafters. They offer a range of projects suitable for varying skill levels and interests, which not only allows you to expand your reach into a diverse range of viewers, but also retains interest as your audience grows their skills.

DIY GARDENING

Ever thought of starting your own garden? If so, this video idea collection will fit perfectly, because DIY gardening challenges cover everything from the basics of urban gardening to advanced rural horticulture.

It doesn't matter whether you are an expert or all your houseplants die no matter what you try; even sharing your journey of *attempting* to build your gardening skills will be fascinating to watch. It's all about creating fun, quirky spaces and learning—or showing—how to build sustainable, beautiful green areas in and around your own home.

Hands-on activities, guides, or even relatable rants about the challenges that you face when building your own garden will have viewers binge-watching you in no time.

So, get your green thumbs ready and try these 10 beautifully green ideas to get you started.

172. Vertical Garden Challenge

Search for innovative ways to create a vertical garden. Find tutorials, guides, and inspirational ideas, or think of your own. Gather the materials and tools you'll need, whether that's specialized equipment or repurposed materials. Share the whole experience with your audience, from finding a great idea to the finished product.

Talk about how you're optimizing space, the challenges you face during the crafting and planting stages, and the vision that you have for the final garden. Ask your audience about their feelings when it comes to vertical gardens, and whether they think you've done a good job.

173. Edible Flowers Garden

For this idea, you need to research edible flowers and the benefits that they offer. Choose a variety of these flowers and share your choices and the reasons for them with your audience. Go through the process of planting and explain how you're going to care for this garden as it grows.

Document the growth process right up to the point where you turn the flowers into culinary creations fit for a queen. You can even add a timelapse of the growth process as it progresses. Ask your viewers to share their favorite edible flowers in the comments, and request suggestions for dishes that you can make from any that you've grown.

174. Terrarium Building Tutorial

Learn the art of building beautiful, low-maintenance terrariums and share the journey of gaining knowledge with your viewers. This kind of garden is perfect for those looking to bring greenery indoors, or even those who would like to add small reptiles or amphibians to their homes.

Share what you've learned, the tips and tricks that have worked for you, and the plants that you think are best suited to the terrarium life. Encourage your audience to create something according to their aesthetic and to share their creations on your other social media platforms. This will create a cross-platform community and improve your connection with your audience.

175. Seasonal Plant Propagation

Discover the secrets of propagating plants from seeds, cuttings, or divisions, with a focus on different techniques for each season. Share your exploration with your viewers, along with all of the triumphs and failures that you experience. Explain the choices you've made for the next season when it comes to plants, and share tips on what each plant might need.

Encourage your audience to follow along for the whole year as you grow different plants in your garden and create entirely new aesthetics every few months. This can be a long-form series that will offer binge-worthy content for your viewers and keep them coming back for more.

176. Fairy Garden Design

Dive into the magical world of fairy gardens, exploring miniature landscapes and whimsical decorations for a touch of enchantment. Search for inspiration across the internet and share your ideas with your audience, even before you start *actually* crafting your fairy garden.

Find the perfect spot, whether inside your house or outdoors and gather the materials you'll need. Bring your audience along as you craft your miniature world of fairies and showcase the final product when placed in the garden.

177. Herb Spiral Construction

Document the process of learning how to design and build your own functional herb spiral to optimize space and add an aesthetic touch to your garden. Showcase the hacks that you use to personalize the spiral to your home and offer tips and tricks to your audience as you go along.

Film the growth of your herbs and use a time-lapse in your video to show their progress. Ask your audience which herbs they think you should add to the spiral next and what dishes they think you should make with them.

178. Indoor Plant Care Tips

Search for expert advice or have an interview with a horticultural master in which you discuss caring for indoor plants, covering topics like proper watering, light conditions, and troubleshooting common issues. This will improve the credibility of your channel and ensure that viewers know they can trust you for the best advice.

Ask your audience for suggestions on more experts that they'd like to see as guests on the channel, and encourage them to share their opinions in the comments.

179. Bonsai Tree Shaping

Delve into the art of bonsai, exploring techniques for shaping and maintaining these miniature trees to create living works of art. Showcase your own bonsai and share interesting facts that you've found online. You can even create reaction content by looking at incredible bonsai shapes that others have made.

Discuss the time that it takes to truly grow a great bonsai tree as well as the dedication that these plants require. Ask your audience to share their experiences with these miniature trees and to give their input on what they think a great shape would be to grow.

180. Pollinator-Friendly Garden Design

Explore ways to design a garden that attracts and supports pollinators, such as bees and butterflies, cultivating a healthy ecosystem. Research the importance of pollinators and share the knowledge you gain with your viewers, adding fun facts and learning moments into your video.

You can even clip these fun facts to add to short-form content on other platforms. Encourage your audience to take part in the discussion in the comments and on your other social media to raise awareness.

181. Succulent Arrangement Challenge

Take on the challenge of creating unique and visually appealing succulent arrangements, experimenting with different varieties and container choices. Choose different events and themes and create succulent arrangements for each one. For instance, you can create arrangements suitable for a wedding or arrangements reminiscent of a particular desert somewhere in the world.

Ask your audience to choose their favorite arrangement and to offer their critiques in the comments. Create polls in the YouTube community section so that your viewers can vote on the best arrangement that you've made.

These DIY gardening video themes cater to a wide range of gardening interests, from beginners all the way to seasoned enthusiasts. They offer your viewers both inspiration and practical tips for cultivating green spaces. So, get out there and make the world a greener place!

DIY Home Renovation

DIY home renovation is a staple of content creation—everyone dreams of the perfect home with the perfect aesthetic. As a creator, your viewers can live vicariously through you and feel as if they're part of the process of turning a house into a home.

You're not simply fixing leaky faucets or unclogging a drain; instead, you're reinventing spaces to suit your own style. This genre also offers opportunities for more long-form content, or even a series, where you can tackle a single room, or even your entire house, step-by-step.

There are so many creative opportunities where you can offer your viewers advice on affordable options or on ways to make spaces feel bigger or more open. You can share your self-expression and encourage your viewers to do the same, creating a strong and uplifting community in the process.

Really, home renovation is a brilliant way to connect with your audience. So, grab your tool belt and start creating with the following 7 ideas.

182. Room Makeover on a Budget

Document the process of transforming a room while keeping yourself limited to a specific budget. Show your viewers how they can use creative solutions and cost-effective materials to achieve a stylish makeover without breaking the bank. You can create a series by starting with a bathroom and moving on to every other room in the house on each episode.

Or, you can showcase different methods and tips to achieve different aesthetics with the same limited budget. Encourage your viewers to leave suggestions for room themes and projects that you can take on in the comments.

____

____

____

____

183. Smart Home Technology Integration

Search the internet for the most high-tech or strange forms of smart technology that you can find. Bring your audience along on the journey as you integrate this technology into your home, documenting everything from the installation of devices to how they actually work. From security systems to automated lighting and thermostats, there are a wide range out there to try.

You can share your opinions and reviews for each device and explain why you think it might be worth it—or not. Showcase the pieces of smart technology that you *actually* find useful in your daily life, and encourage your viewers to share their opinions on homes getting smarter in the comments.

184. Outdoor Living Space Enhancement

Bring your audience along for the ride as you document improvements to an outdoor area, such as a backyard or a patio. Include DIY projects like building a deck, creating outdoor furniture, or adding landscaping elements. Research the best tips and tricks for optimizing space, and share what you learn with your audience.

Encourage engagement by asking viewers to comment on their favorite outdoor spaces and to share their own tricks for creating the perfect place to relax. Share the final product by hosting a party, a picnic, or another event for friends, family, or other video-makers.

185. Feature Wall Installation

Document yourself as you elevate a room's aesthetics by installing a stunning feature wall, exploring various materials like reclaimed wood, shiplap, or even decorative panels. Showcase the process of designing the wall and gathering the materials to create it, explaining your choices to your audience as you go.

Share the challenges of the process and how you overcame them to turn them into triumphs. Add unique elements to the final design to truly express your own personal style, and ask your audience for their opinions on your new wall in the comments.

186. Basement Transformation

Document the process of converting an unused basement space into a functional and stylish area, whether it's a home office, an entertainment room, or even a guest suite. Showcase how you deal with basement-specific challenges and offer your viewers tips and tricks on creating the perfect space below ground.

Add in unique elements to create a style that is all your own and that makes the space feel fantastic. Ask your viewers what kind of space they'd create in their own basements, or create a poll in the community sections for viewers to vote on.

187. DIY Flooring Upgrade

Document the process of searching for and exploring various DIY-friendly flooring options. Explain your choices and reasoning, and demonstrate the process of upgrading your floors, whether that's installing laminate or refinishing hardwood, carpeting, or tiles. Add tips for making the process easier, or for making the new floors last longer, and explain the things that should be avoided during the process.

Show off your final product with a bang by doing a dance on your new floors or by hosting a floor party. Ask your audience to share their opinions and questions in the comments.

____

____

____

____

188. Home Office Nook Creation

This can be done in your filming area or in another home office. Design and build an office nook or a study space, emphasizing practicality, organization, and creative design for a productive work environment. Share your design choices and reasoning with your audience, and offer advice on staying productive even when it's difficult.

Ask your audience to comment on whether they'd like to see the new nook in more videos, and whether they'd like more advice on keeping their work environment organized and fun. Encourage discussion in the comments on productivity methods so that your community can help one another out.

These DIY home renovation themes offer many kinds of projects while catering to an audience that might be interested in improving their living spaces with hands-on, budget-friendly solutions. It also offers you, as a creator, a place to create community around improving your home and is a great avenue for self-expression. So, get your tools ready, and get moving!

Reverse Engineering

Reverse engineering is a fascinating adventure. In these videos, you, as a creator, dive into the deep end and investigate the inner workings of modern technological devices. Through taking gadgets and doodads apart, you discover exactly how they're put together —and *why*. This gives you the opportunity to expose the truth behind the mysteries of tech by teaching your audience all about how things are made and how they work.

You can also reverse engineer software, furniture, toys—really, anything that is man-made and put together from multiple parts. Viewers are fascinated with unique knowledge, and it's an opportunity that you shouldn't let slip through your fingers.

Grab your tools and get going; 9 high-tech ideas await you below.

189. Gadget Teardowns

Set up your equipment and film the process as you disassemble and explore the inner workings of popular gadgets, from smartphones and laptops to gaming consoles and smartwatches. You can even disassemble more bizarre inventions or rarely used tech for an interesting twist.

Go into detail about each component and its role in the device you're displaying and add interesting facts about the manufacturing process. Ask your audience for suggestions on the next device to disassemble and create a fascinating series of content that will keep them watching for hours.

190. Automotive Component Analysis

Break down and analyze various automotive components, like engines, transmissions, or electronic systems, to help your viewers understand their functionality and design. This doesn't have to be limited to cars; you can use 3-D modeling and editing techniques to dive into larger and smaller vehicles of any kind, from bicycles all the way through to jet engines and even the international space station.

Offer your viewers insight and fun facts about various vehicles, and ask them which vehicle's inner workings they'd like to see you talk about next. You can clip smaller sections of your video to create short-form content for other platforms, as well, to bring more viewers to your channel.

191. Appliance Dissection

In this video, you'll film the process of taking apart household appliances such as microwaves, blenders, or washing machines, revealing the remarkable engineering behind everyday conveniences. If you don't have access to enough appliances to pull apart and inspect, you can also use your 3-D modeling skills to take them apart virtually.

Go into depth when dissecting these appliances, showing your audience how everything works down to the smallest components. For an extra challenge, you can attempt to reassemble the appliance at the end of the video!

192. Electronic Circuit Reverse Engineering

Dive into the creation of circuit boards by reverse engineering them, exploring their differences, similarities, design, and functionality. Use the best camera you can to zoom in closely so that your audience can see even the most minute details of these electronic wonders.

Discuss each component and its importance, as well as what would happen if that component was not part of the design. You might even want to go into the history of circuit boards—how they were first invented and how they have evolved over time. Ask your audience to comment on the type of circuits they're most curious about, and use that information as inspiration for your next video.

193. Software Deconstruction

Analyze and reverse engineer software applications, exploring coding techniques, algorithms, and security measures that have been implemented in these programs. Use screen recordings to bring your audience along for the journey, showing exactly how the most common applications used today work behind the scenes. Make sure that you know what you're allowed to show and not to show, so you don't cause any trouble.

While you're doing this, you can also offer reviews of various programs as well as options and alternatives that your viewers can look into. Compare pros and cons, and ask your audience about their favorite applications.

194. Toy Deconstruction

Disassemble and inspect the inner workings of toys, showcasing the engineering behind them and how they achieve specific functions. Explain your opinion on various design choices and safety features, and go into detail when discussing the various components.

Use camera angles, zoom, and editing software to ensure that your audience can truly absorb every step and join you in your journey of discovery. Ask viewers to discuss their favorite childhood toys in the comments, and get inspiration from their suggestions for your next video.

195. Architectural Design Breakdown

In these videos, you strive to explore architectural designs through reverse engineering, breaking down construction techniques and the structural elements of buildings and structures. You can choose famous places like the Golden Gate Bridge or the Empire State Building or simple local buildings—or even ask for suggestions in the comments for different bizarre structures and buildings that your audience would like to see.

Talk about various design elements and the ways in which engineers likely brought them to life. You can even put a spin on this niche by talking about historical architecture and the differences between older and newer buildings.

196. Industrial Machinery Disassembly

Whether by using 3-D modeling software or diving into the real thing, create a video in which you dissect large-scale industrial machines or equipment. As you go down to the smallest components, uncover the intricate mechanisms and engineering solutions used by the machines' builders.

Talk about the history of manufacturing and mention fun facts as you go along, bringing your viewers on a journey of fascinating knowledge. Discuss the impact of machines on manufacturing and how it has changed the landscape of industry over the last few hundred years, which will add depth to your video.

197. Product Restoration and Reconstruction

Reverse engineer old or damaged products, showing your audience each step of the restoration process as you bring new life to these items. Describe what these products were used for and the impact they had on the world in different time periods. Showcase the materials and specialized tools that might be necessary for any number of restoration projects.

Ask your viewers for input on more products that you can restore, and encourage them to share stories about strange or unique products from their youth, or from previous generations.

These reverse-engineering video themes offer viewers insights into the engineering and design aspects of various products from consumer electronics to industrial machinery. They allow you to share your knowledge and foster a deeper understanding of the world around us. Why not go ahead and help your audience learn something new?

Fusion Cooking DIY Experiments

Fusion cooking DIY experiments are art in the form of food. They go against the norms of the culinary world and bring a whole wealth of new flavors into your kitchen. You might want to try out exotic ingredients or even stay close to home; whatever you choose, the most important thing when it comes to fusion cooking is the *experiment.*

In these DIY challenges, the idea is to combine the foods of various cultures into something beautiful with a new, unique flavor. Viewers will follow along to see what creative ideas you can bring to the table next and which innovative dishes you will create. Don't be afraid to try something new—you might be surprised!

The 8 ideas that follow can help get you cooking.

198. Sushi-Burrito Fusion

In this experiment, you will combine the flavors of Japanese sushi and the convenience of the Mexican burrito. Delight your audience as you experiment with different sushi fillings and rolling techniques to create the ultimate sushi-rito. Test out the various results of your experiment and give a review for each one.

Encourage your audience to give their opinions on this unique fusion in the comments, offering their ideas for fillings and imagining the various flavor possibilities.

199. Mexican–Italian Pasta Tacos

Create a video where you infuse the flavors of traditional Italian pasta dishes into Mexican-style tacos, building a fusion of two beloved, rich cuisines. Try out various combinations of spice and sauce to create the perfect pasta taco, exploring the originating cultures as you go.

You can add in fun facts about Mexican and Italian cuisine for a bit of an educational spin on the topic, and taste-test your final creations with a fun dinner party. Ask your viewers to share their opinions on your creations in the comments to foster discussion and connection.

200. Indian–Chinese Stir-Fry Paratha

In this challenge, your aim is to perfectly mix the spices of Indian cuisine with the savory stir-fry concept of Chinese cooking, wrapping it all up in a traditional Indian paratha. Discuss the various nuances of cuisine in each country and the flavors that you think will blend together the best.

Create several options for yourself to try, explaining your choices and decisions to your audience as you go along. At the very end, test the final products and rank them in a tier list from best to worst, encouraging your audience to add their opinions to the mix in the comment section, maybe even ranking for you as well.

201. Mediterranean Sushi Rolls

Experiment by choosing a range of Mediterranean flavors, and create sushi rolls with ingredients like hummus, olives, and feta cheese. Mix these with different types of sushi and sauces to create the perfect flavor-rich combination. Discuss the pairings of flavors that you choose with your audience and the reasons behind them.

Ask your audience to vote on which combination looks and sounds the most delicious, and test each one out in a fun taste-testing challenge.

202. Thai–Indian Curry Pizza

In this challenge, your aim is to merge the aromatic flavors of Thai and Indian curry with the comfort of traditional Italian pizza. Showcase your culinary talent by bringing your viewers along as you experiment with curry-infused sauces and toppings to find the perfect combination.

Offer close-ups and taste tests of each combination, making sure to use a palette cleanser in between so that the different spices don't overwhelm one another. Give a review of each blend and announce the best of them at the end of your video.

203. Greek–Japanese Gyro Sushi Bowl

In this experiment, you're going to combine elements of a Greek gyro with the presentation of a Japanese sushi bowl. Combine gyro meat, tzatziki, and fresh veggies to create a beautiful and delicious fusion. You can also host a themed dinner to celebrate both cultures along with your friends.

Add in some facts about Japanese and Greek cuisine and the reasons that certain flavors go well with others. You might even want to add in a few wine pairings to make the meal extra special.

204. Hawaiian–Indian Fusion Poke Bowl

Blend tropical flavor and Eastern spice in this experiment, in which you combine the richness of Hawaiian poke with the beautiful heat of Indian cuisine, creating a truly unique combination. Bring together friends for a picnic and have them test the tastes that come from your creation.

Ask them to share their critiques and opinions with your audience, and encourage viewers to do the same in your comments section.

205. Peruvian–Mexican Ceviche Tostadas

Merge the citrusy flavors of Peruvian ceviche with the crunchy texture of Mexican tostadas. Make several different iterations while filming the whole process, experimenting with different toppings and ingredients. Lay out the final set of dishes for your audience, explaining your choices and blends for each one.

At the end of the video, offer a taste test of each flavor combination, noting your own critiques and reviews. Ask your audience which one they'd most like to try, and what they think you should combine next.

These fusion-cooking DIY experiment themes offer a delightful exploration of diverse culinary traditions, encouraging creativity and innovation in the kitchen. They're sure to pull you out of a slump if you've been struggling to find fresh ideas, and they will definitely encourage your audience to try out their own experiments. So put on that chef's hat and get cooking!

DIY Costume Creations

DIY costumes are the ultimate way for you to share your personal self-expression and to put yourself in the spotlight. You can keep within a budget by transforming thrift-store finds, or go all out for an intricately made cosplay idea.

Costume creation is something that's always worked on YouTube, and it will captivate your viewers every time. It's a way to celebrate DIY culture with your own unique flair. Show off the process of choosing materials, designing, and creating, and finally reveal the result of your hard work—and don't forget to include the challenges that you faced along the way.

Get ready to create a community where you inspire your viewers and they inspire you through beautiful fashion and through encouragement. Use the following 7 suggestions to get you started.

206. Thrift Store Transformation

Document the journey as you visit various thrift stores, searching for unconventional and unique items that you can transform into impressive costumes. Once you've found your items, take your audience back to your home, where you explain your vision for all of the different pieces.

Have them follow along with your process as you transform your thrift store finds into something incredible, explaining each step and decision as you go. You can even create multiple costumes and compile the final reveals into clips for short-form content platforms, which can help you build a wider reach.

207. Paper Couture Challenge

Gather other video-makers and challenge them to craft elaborate costumes using only paper materials, illustrating the versatility and ingenuity of paper in the world of fashion. If you want the challenge to be more difficult, you can also add a time limit and livestream it for viewers to enjoy. Once everyone's creations are complete, hold a physical or virtual fashion show to showcase the final looks.

Create a poll in the live stream or in the community section for your audience to vote on their favorite looks. Clip together moments from the live stream to create more content for your channel or for short-form platforms.

208. Food-Inspired Fashion Show

Combine the worlds of food and fashion by bringing together creators and challenging them to join you in designing costumes inspired by various dishes. You can take a more serious and dramatic approach by creating runway-worthy avant-garde attire, or make it more humorous by creating caricatures or more cartoon-like outfits. You can even combine the two to truly capture your audience's imaginations.

Have various creators come together to discuss their creations and the inspirations behind the choices that they made. You can even add in clips from their different creation processes to show the diversity in artistic expression.

209. Invisible Costume Challenge

Gather other creators and challenge them to design costumes playing with the idea of invisibility. The idea is to use and show your audience creative techniques and optical illusions that can make various parts of their bodies appear "invisible." There are a variety of ways to achieve this, and the choices that each creator makes will illustrate the unique nature of personal expression and how different people approach solving problems.

Ask your audience to confirm whether they were fooled by any of the final products and ask them what kind of challenge you can offer them next.

210. Future Fashion Forecast

Document yourself as you analyze trends of the past and the present and gather other creators together in a challenge in which participants have to predict the trends of the future. Showcase unique personalities and ways of thinking by asking creators to craft costumes representing their personal vision of futuristic styles, exploring innovative materials and concepts.

You can add a bit of humor into the mix by asking other video-makers to create costumes that they envision aliens from other worlds wearing as tourists to Earth, and vice versa. Ask your audience to share their opinions on what they think the future of fashion looks like in the comments.

211. Outer Space Extravaganza

Challenge yourself to design and craft various costumes inspired by outer space, exploring various cosmic themes to create otherworldly attire. You can incorporate alien fashion for various creatures inspired by Sci-Fi, or go down the route of imagining what each of our solar system's planets would wear if they were human for a day.

There are a wide variety of options, since space is quite literally limitless. You can also challenge other video-makers to take part in this challenge and host a virtual space-themed fashion show.

____

____

____

____

212. DIY Pet and Owner Costume Challenge

If you have pets, this can be an adorable and truly wholesome idea to implement. Document the process of creating matching costumes for yourself and your pets, being sure to include all of the challenges and triumphs. You can even create costumes for friends and family and their pets, and then have a get-together where everyone gets to show them off.

Try out various themes or character combinations, or even go for something seasonal or holiday specific. Ask your viewers to vote for their favorite pet-and-owner costume, and which ideas they'd like to incorporate in their next special event.

These DIY costume challenge themes offer a captivating blend of creativity, innovation, and craftsmanship, making them enjoyable for viewers who appreciate the art of handmade fashion. They also offer a wide range of ideas-within-ideas, ensuring that it will be difficult for you to run out of content.

So, What Now?

In these pages, you've found a valuable resource, a treasure trove of inspiration for entertainers and creatives looking to carve a niche in the endless world of content creation. From the very first chapter, we looked at the art of crafting captivating YouTube content by going through a menagerie of viral ideas that are tailored to spark your imagination.

In Part One: YouTube Ideas for Entertainers, we discovered the power of laughter in creating a dynamic and powerful community that is geared toward growth. Here, you found everything from the best blindfolded challenges to obstacle course ideas that will have your audience captivated from the moment you hit Record.

As an entertainer, these ideas are geared to equip you with the tools and knowledge that you need to turn wild concepts into viral sensations. Hopefully, you've found exactly what you needed to bring your personal mojo back and to keep entertaining for a long time.

In Part Two: YouTube Ideas for Creatives, we dove into the human imagination, exploring the limitless avenue of creative endeavors for your audience to enjoy. From deep how-to guides to attempting your own garden, we discussed ideas that will empower you to find your voice and express yourself through your content.

These ideas are designed to keep audiences watching throughout their crafting and building journey, from beginners to more advanced users. We covered educational content, content geared toward expression, and content that will build their network with other creators and take them to the next level.

Beyond the strategies and tactics that this book has offered you, there lies a deeper truth: At the very heart of every viral idea lies the power to connect, inspire, and entertain. Beyond the metrics of likes and views, you have the power to spark meaningful conversation, foster engagement, and even create a positive change in the world.

After all, behind every screen is a human being craving connection and moments of joy, and as video-makers, it's our responsibility to deliver them.

Don't be afraid to dive into the deep end and to try something new. Remember to enjoy yourself, and to bring your own unique personality to every video. Get out there and share yourself with the world—and, if you need more ideas, don't be afraid to come back!

Thank You

Thank you for purchasing this book and supporting my endeavors to share knowledge. I would really appreciate it if you could leave a review and let me know what you think!

Also By

Ben Davis

Some other books that Ben has written are:

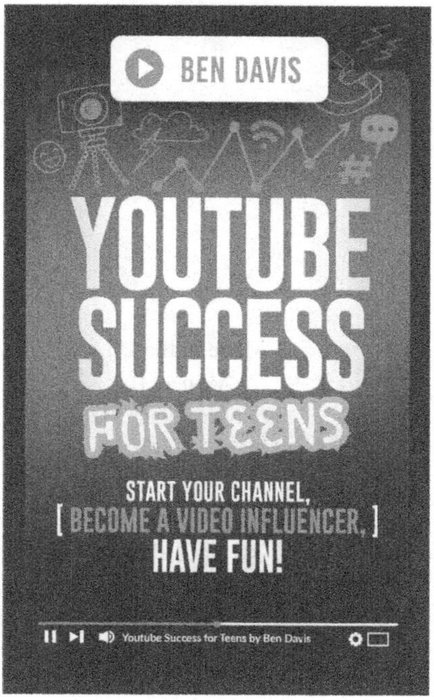

Made in the USA
Monee, IL
04 November 2026

67377196R00105